Diana Langer

Asymmetric Cell Division in Drosophila melanogaster Neuroblasts

Diana Langer

Asymmetric Cell Division in Drosophila melanogaster Neuroblasts

Identification of Miranda Associated Proteins and RNA in Drosophila melanogaster Neuroblasts

Südwestdeutscher Verlag für Hochschulschriften

Impressum/Imprint (nur für Deutschland/ only for Germany)
Bibliografische Information der Deutschen Nationalbibliothek: Die Deutsche Nationalbibliothek verzeichnet diese Publikation in der Deutschen Nationalbibliografie; detaillierte bibliografische Daten sind im Internet über http://dnb.d-nb.de abrufbar.

Alle in diesem Buch genannten Marken und Produktnamen unterliegen warenzeichen-, marken- oder patentrechtlichem Schutz bzw. sind Warenzeichen oder eingetragene Warenzeichen der jeweiligen Inhaber. Die Wiedergabe von Marken, Produktnamen, Gebrauchsnamen, Handelsnamen, Warenbezeichnungen u.s.w. in diesem Werk berechtigt auch ohne besondere Kennzeichnung nicht zu der Annahme, dass solche Namen im Sinne der Warenzeichen- und Markenschutzgesetzgebung als frei zu betrachten wären und daher von jedermann benutzt werden dürften.

Verlag: Südwestdeutscher Verlag für Hochschulschriften Aktiengesellschaft & Co. KG
Dudweiler Landstr. 99, 66123 Saarbrücken, Deutschland
Telefon +49 681 37 20 271-1, Telefax +49 681 37 20 271-0
Email: info@svh-verlag.de
Zugl.: München, Ludwig-Maximilians-Universität München, Dissertation, 2008

Herstellung in Deutschland:
Schaltungsdienst Lange o.H.G., Berlin
Books on Demand GmbH, Norderstedt
Reha GmbH, Saarbrücken
Amazon Distribution GmbH, Leipzig
ISBN: 978-3-8381-1247-3

Imprint (only for USA, GB)
Bibliographic information published by the Deutsche Nationalbibliothek: The Deutsche Nationalbibliothek lists this publication in the Deutsche Nationalbibliografie; detailed bibliographic data are available in the Internet at http://dnb.d-nb.de.

Any brand names and product names mentioned in this book are subject to trademark, brand or patent protection and are trademarks or registered trademarks of their respective holders. The use of brand names, product names, common names, trade names, product descriptions etc. even without a particular marking in this works is in no way to be construed to mean that such names may be regarded as unrestricted in respect of trademark and brand protection legislation and could thus be used by anyone.

Publisher: Südwestdeutscher Verlag für Hochschulschriften Aktiengesellschaft & Co. KG
Dudweiler Landstr. 99, 66123 Saarbrücken, Germany
Phone +49 681 37 20 271-1, Fax +49 681 37 20 271-0
Email: info@svh-verlag.de

Printed in the U.S.A.
Printed in the U.K. by (see last page)
ISBN: 978-3-8381-1247-3

Copyright © 2010 by the author and Südwestdeutscher Verlag für Hochschulschriften Aktiengesellschaft & Co. KG and licensors
All rights reserved. Saarbrücken 2010

TABLE OF CONTENTS

Summary

1. Introduction

    1.1. Asymmetric Cell Division   1
        1.1.1. General Aspects of Asymmetric Cell Division   1
        1.1.2. Asymmetric Cell Division in *Drosophila melanogaster*   2
    1.2. *Drosophila* Neuroblasts as Model to Study Asymmetric Cell Division   3
        1.2.1. Neurogenesis in *Drosophila melanogaster*   3
        1.2.2. The Asymmetric RNA/Protein Localization Network in Neuroblasts   6
        1.2.3. Role of Cell Cycle Regulators in Neuroblast Cell Divisions   12
            1.2.3.1. Cdc2   12
            1.2.3.2. Aurora A and Polo Kinases   12
            1.2.3.3. Cyclin E   13
            1.2.3.4. The Anaphase Promoting Complex/ Cyclosome   15
        1.2.4. Starting and Stopping Neuroblast Divisions   15
    1.3. RNA Localization   18
        1.3.1. General Role of RNA Localization   18
        1.3.2. Mechanism of mRNA Localization   19
        1.3.3. Staufen has a Conserved Role in RNA Localization   20
        1.3.4. Staufen Dependent RNA Localization in *Drosophila*   21
    1.4. Goals of the Thesis   23

2. Materials and Methods

    2.1. Materials   24
        2.1.1. Chemicals   24
        2.1.2. Enzymes   24
        2.1.3. Kits   24
        2.1.4. Antibodies   25
            2.1.4.1. Commercially available Antibodies   25
            2.1.4.2. Non-commercial Antibodies   25
        2.1.5. Fly Stocks   26

2.2. Methods	26
2.2.1. Standard Laboratory Methods for *Drosophila melanogaster*	26
2.2.1.1. Laboratory Culture	26
2.2.1.2. Embryo Collection	26
2.2.2. Methods in Molecular Biology	27
2.2.2.1. Oligonucleotides	27
2.2.2.2. Preparation of Digoxigenin Labeled RNA Probes	33
2.2.2.3. RNA Preparation and Reverse Transcription	34
2.2.2.4. Candidate PCR Analyses	35
2.2.3. Methods in Biochemistry	36
2.2.3.1. Sypro Ruby Protein Staining	36
2.2.3.2. Preparation of GST-Miranda Beads	
2.2.3.2.1. Protein Expression	36
2.2.3.2.2. Protein Purification and Preparation of Beads	37
2.2.3.3. GST Pull-Down Experiments	37
2.2.3.3.1. Preparation of *Drosophila* Embryo Extract	37
2.2.3.3.2. GST Pull-Down	38
2.2.3.4. Immunoprecipitation Experiments	38
2.2.3.4.1. Preparation of *Drosophila* Embryo Extract	38
2.2.3.4.2. Immunoprecipitation and Westernblot	39
2.2.3.5. Sucrose Gradient	40
2.2.3.6. Gelfiltration	41
2.2.4. Immunostaining and *In situ* Hybridization	41
2.2.4.1. *Drosophila* Embryo Staining	41
2.2.4.2. *In situ* Hybridization	42

3. Results

3.1. Identification of Novel Miranda Protein Interaction Partners	44
3.1.1. Expression and Purification of GST-Miranda	44
3.1.2. GST Pull-Down Experiments	45
3.1.3. GST Pull-Down Candidate Analyses	48
3.1.3.1. Tudor-SN	48
3.1.3.2. Headcase	50

3.1.4. Immunoprecipitation Experiments	51
3.1.5. Pavarotti Analyses	54
3.2. Biochemical Characterization of Miranda Complexes	55
3.2.1. Linear 10%-50% Sucrose Gradient	55
3.2.2. Gelfiltration	57
3.3. Identification of Novel RNAs, Associated to Miranda Complexes	58
3.3.1. Miranda Immunoprecipitation and Candidate PCR Analyses	58
3.3.2. *Dacapo In situ* Hybridization Experiments	64
3.3.3. Size Quantification of *Dacapo*/ Miranda Co-Expressing Neuroblasts	69
3.3.4. *Dacapo* RNA and Protein Staining	70
3.3.5. Dacapo Mutant Analyses	71
3.3.5.1. BrdU Labeling	71
3.3.5.2. Caspase-3 Staining	72
3.3.5.3. Dacapo Mutant Analyses in the Neuroblast 6-4 Lineage	74

4. Discussion

4.1. Identification of Novel Miranda Protein Binding Partners	77
4.2. Biochemical Characterization of Miranda Complexes	82
4.3. Identification of Novel RNAs, Associated to Miranda Complexes	83
4.4. Conclusion and Outlook	87

References

Summary

Asymmetric cell divisions generate cell diversity. *Drosophila* neuroblasts divide in an asymmetric manner to generate another neuroblast and a differentiating cell, namely the ganglion mother cell. The adaptor protein Miranda plays a crucial role in creating intrinsic differences in the daughter cells, by asymmetrically localizing key differentiation factors.

This thesis describes the identification of further partners of Miranda and investigates the existence of Miranda containing complexes.

In fact, GST pull-down and immunoprecipitation experiments could identify Tudor-SN and Headcase as Miranda partners. They seem to bind transiently and most likely do not participate in Miranda's localization. Sucrose gradient and gelfiltration experiments reveal the existence of at least two Miranda containing complexes. One complex with an approximate size of 660 kDa does not show any sensitivity to RNAse treatment. The second with the approximate size of at least 2 MDa, exhibits RNAse sensitivity. Interestingly, an additional RNA that is asymmetrically segregated to the ganglion mother cell could be identified. The RNA corresponds to Dacapo, the *Drosophila* CIP/KIP-type cyclin dependent kinase inhibitor, specific for Cyclin E/ Cdk2 complexes. This result confirms the importance of Miranda in RNA localization in *Drosophila* neuroblasts.

Altogether, the performed experiments provide a starting point for further investigations on the role of the versatile and multi-functional Miranda protein not only in neuroblast divisions, but probably in other cellular processes that require RNA transport in *Drosophila*.

1. Introduction

1.1. Asymmetric Cell Division

1.1.1. General Aspects of Asymmetric Cell Division

Asymmetric cell division (ACD) is a conserved process required to generate cell fate diversity. This type of division results in two distinct daughter cells, in contrast to normal cell divisions which give rise to equivalent daughter cells.

ACD can be achieved by either extrinsic or intrinsic mechanisms. Extrinsic mechanisms require cell signalling events between cells (Morrison et al, 1997). Intrinsic mechanisms involve the preferential segregation of cell fate determinants to one of two daughter cells during mitosis. A prerequisite for the asymmetric segregation of cell fate determinants is that the mother cell has to be polarized and the mitotic spindle has to be aligned with the axis of polarity (reviewed in (Kaltschmidt & Brand, 2002)).

Asymmetric divisions often give rise to only one novel cell type in addition to a new copy of the mother cell. Self renewal is a feature of stem cells and there exists growing evidence that stem cells self-renew through asymmetric divisions (Macieira-Coelho, 2007). ACD have been well characterized in mouse, the nematode *Caenorhabditis elegans* and the fruit fly *Drosophila melanogaster* (reviewed in (Betschinger & Knoblich, 2004).

Asymmetric divisions have recently been shown to regulate cell fate decisions in the mammalian haematopoietic system. It represents one of the best understood stem cell lineages in mammals. Hematopoietic stem cells give rise to all types of blood cells. They were shown to be able to divide not only symmetrically but also asymmetrically. The direct mechanism of the asymmetric cell division is not clearly understood, but apparently different levels of Notch signalling in the two daughter cells play a role (Schroeder, 2007; Wu et al, 2007).

1.1.2. Asymmetric Cell Division in *Drosophila melanogaster*

Drosophila melanogaster harbours several cell types which show asymmetric features. Among these are the germline stem cells (GSCs). In each germarium (region of the ovary that contains the stem cells), 2-3 GSCs are surrounded by an equal number of cap cells, which form the stem cell niche. They are connected by adherens junctions and their removal results in stem cell loss. This suggests that niche adhesion is essential for GSC maintenance (Song et al, 2002).

Recently, other stem cell lineages, showing asymmetric divisions have been discovered in the fruit fly. They are found in the adult gut (Ohlstein & Spradling, 2006; Ohlstein & Spradling, 2007), in the malphigian tubules (Affolter & Barde, 2007; Micchelli & Perrimon, 2006; Singh et al, 2007) and in the haematopoietic system (Krzemien et al, 2007; Mandal et al, 2007).

Two types of well characterized asymmetrically dividing precursor cells are found in the developing *Drosophila* nervous system. Sensory organ precursor (SOP) cells represent the neural precursor cells of the peripheral nervous system (PNS), whereas neuroblasts are the precursor cells of the central nervous system (CNS).

SOP cells give rise to the four cells types present in external sensory organs, which are the socket, the hair, the sheath and the neuron cells (Figure 1).

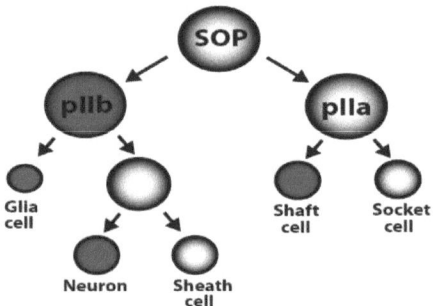

Figure 1. The *Drosophila* Sensory Organ Precursor Lineage.
The SOP cell divides into an anterior pIIb and a posterior pIIa cell. These cells differentiate further into a neuron, a sheath cell, a shaft cell and a socket cell.

Cells within the lineage that inherit the cell fate determinant Numb are marked in pink. The same lineage was described in the embryonic peripheral nervous system and in the bristle (microchaete) lineage of the adult fly. Cells within the lineage that inherit the cell fate determinant Numb are shown in pink.

After delaminating from a polarized epithelium, the SOP cells divide into an anterior pIIb and a posterior pIIa cell. These two cells then divide once more to generate the two outer and the two inner cells of the organ.

The asymmetry in all of these divisions is established by the different levels of Notch activity in the daughter cells, due to an unequal distribution of the cell fate determinant Numb (Le Borgne et al, 2005; Rhyu et al, 1994; Schweisguth, 2004).

Numb acts as a tissue-specific repressor of the Notch pathway (Le Borgne et al, 2005; Schweisguth, 2004). In *numb* mutants both SOP daughter cells adopt the cell fate of the one that normally does not inherit the protein. In accordance with that, *numb* overexpression results in a transformation of both cells to the same fate.

1.2. *Drosophila* Neuroblasts as Model to Study Asymmetric Cell Division

1.2.1. Neurogenesis in *Drosophila melanogaster*

Besides the SOP cells in the PNS, the *Drosophila* nervous system also harbours neural precursors of the CNS, the neuroblasts (NB).

About 30 NBs delaminate from the neuroectoderm per thoracic and abdominal hemisegment (Broadus et al, 1995; Doe, 1992). The remaining cells of the neurogenic region remain superficial and generate the ventral epidermis.

The "proneural" genes (Ghysen & Dambly-Chaudiere, 1989) control the position and time at which groups of neuroectodermal cells become competent to form a neuroblast, whereas the "neurogenic" genes (Lehmann R., 1983) control the cell interactions that prevent more than one cell in the group from developing into a neuroblast. One can say that the proneural genes act to neuralize a group of otherwise epidermal cells, whereas the neurogenic genes assure that only one cell within the patch becomes a neuroblast.

The proneural genes include the *achaete-scute* complex (AS-C) with *achaete* (*ac*), *scute* (*sc*), *lethal of scute* (*l'sc*), and *asense* (*ase*).

The AS-C is activated in proneural clusters in the ventral portion of the fly. The proneural genes are expressed in each of the 14 segments, which are defined by the pair rule genes. A further proneural gene, *atonal* (ato), was isolated more recently in a PCR screening for bHLH sequences related to that found in *achaete-scute* complex genes (Jarman et al, 1993). Interestingly, members of the asc and ato families account for all proneural activity in the PNS, but not in the CNS, where the generation of some neuroblasts does not require any of the known proneural genes (Jimenez & Campos-Ortega, 1990).

From each cluster one neuroblast develops, whereas the remaining epidermal cells of the cluster loose proneural protein expression. Once a cell begins to differentiate as a neuroblast, it prevents the adjacent cells from becoming neuroblasts by lateral inhibition, which is mediated by the neurogenic genes.

Two of the neurogenic genes, encoded by *notch* and *delta*, interact directly at the membranes of adjacent cells (Fehon et al, 1990; Lieber et al, 1992), transmitting a signal from the neuroblast to the neighbouring cells that inhibits neural development (Doe & Goodman, 1985; Stuttem & Campos-Ortega, 1991; Taghert et al, 1984).

In *Drosophila* two temporally and in part genetically different types of neuroblasts can be found. These are the neuroblasts of the embryo and of the larvae. Laval neuroblasts generate the thousands of neurons found in the central nervous system of the adult fly.

While embryonic neuroblasts become smaller after each division, larval neuroblasts grow back to their original size after each division and can divide hundreds of times. The two resulting daughter cells (neuroblast and GMC) have nearly equal sizes (Ito & Hotta, 1992; White & Kankel, 1978).

The delamination of embryonic neuroblasts occurs in 5 waves (Figure 2) between embryonic stages 8 and 11 (staging according to (Campos-Ortega, 1985)) (Figure 6). Each neuroblast can be identified by its unique gene expression profile, its time and place of birth and its neuronal and glia progeny.

Each neuroblast has been assigned a name based on a coordinate-like system that relates every neuroblast to its position per hemisegment (e.g. neuroblast NB6-4 is located in row 6 and column 4) (Figure 2).

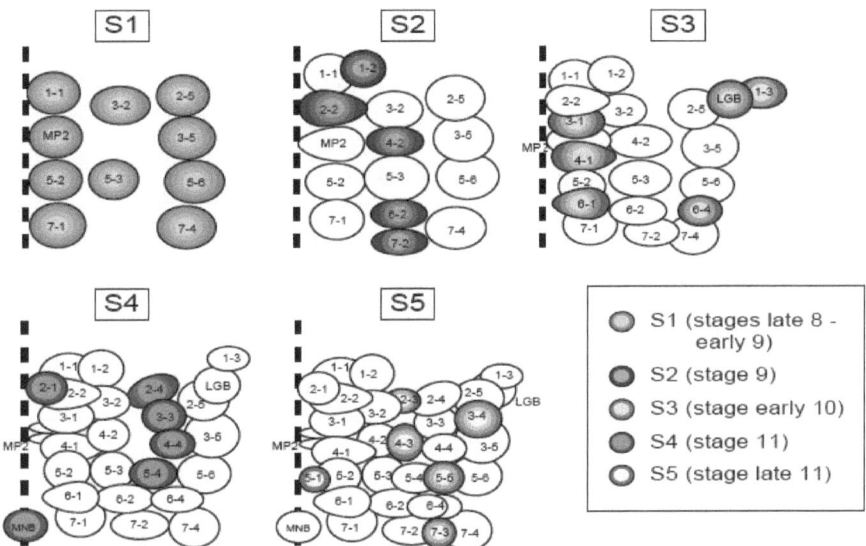

Figure 2. Spatial arrangement and temporal sequence (S1-S5) of segregating neuroblasts. Each map represents the pattern of one hemisegment (thorax, abdomen) with those neuroblasts highlighted that are added during the respective wave of segregation. Ventral midline is marked by broken line (Bossing et al, 1996; Doe, 1992).

Neuroblasts divide asymmetrically in a stem-cell like fashion to produce an apical daughter cell, which remains a neuroblast and a smaller basal intermediate progenitor daughter cell, called the ganglion mother cell (GMC). The GMC then divides once more, to generate either two neurons, one neuron and one glia cell, or 2 glia cells.

In each thoracic and abdominal hemisegment, about 30 NBs delaminate from the ventral neurogenic region.

In total, these 30 NBs produce about 350 progeny cells (30 glia cells, 30 motoneurons and about 290 interneurons) (Bossing et al, 1996; Ito K., 1995; Landgraf et al, 1997; Schmidt et al, 1997).

Every neuroblast produces a near invariant number of neuronal and glia cells (reviewed in (Skeath & Thor, 2003)). Within a given hemisegment, the size of neuroblast clones produced during the embryonic phase of neurogenesis varies immensely. At one extreme, the neuroblast MP2 generates only two cells (Bossing et al, 1996), whereas neuroblast NB7-1 can produce more than 40 cells (Schmid et al, 1999).

1.2.2. The Asymmetric Protein/ RNA Localization Network in Neuroblasts

The key proteins that play crucial roles in setting up neuroblast polarity, which is the essential first step in asymmetric cell division, have been identified (reviewed in (Bardin et al, 2004; Betschinger & Knoblich, 2004; Wang & Chia, 2005; Wodarz & Huttner, 2003)).

Asymmetric cell divisions of neuroblasts are accompanied by localization of protein complexes and RNA to opposite poles (Figure 3), as well as a programmed rotation of the mitotic spindle.

The evolutionarily conserved Par complex consisting of Bazooka (the fly homolog of C.elegans Par-3), Par-6 and atypical Protein Kinase C (aPKC), co-localize at the apical side of the delaminating neuroblast with the NB specific protein Inscuteable (Insc), leading to an apical-basal polarity at this point.

During mitosis, the Insc/ Par complex establishes an apical crescent and recruits another evolutionarily conserved protein complex consisting of Partner of Inscuteable (Pins) and the heterotrimeric G protein subunit Gαi. This leads to maintenance of apical-basal polarity (Parmentier et al, 2000; Schaefer et al, 2000; Yu et al, 2000).

It is suggested that the two apical signalling pathways have overlapping but different roles in asymmetric NB division (Izumi et al, 2004). While the Pins/Gαl complex is mainly involved in spindle orientation, the Par complex induces the asymmetric localization of cell fate determinants to the opposite, basal side of the cell and their segregation into the basal GMC.

There are two discovered cell fate determining complexes that are asymmetrically localized in the *Drosophila* neuroblast (Figure 3).

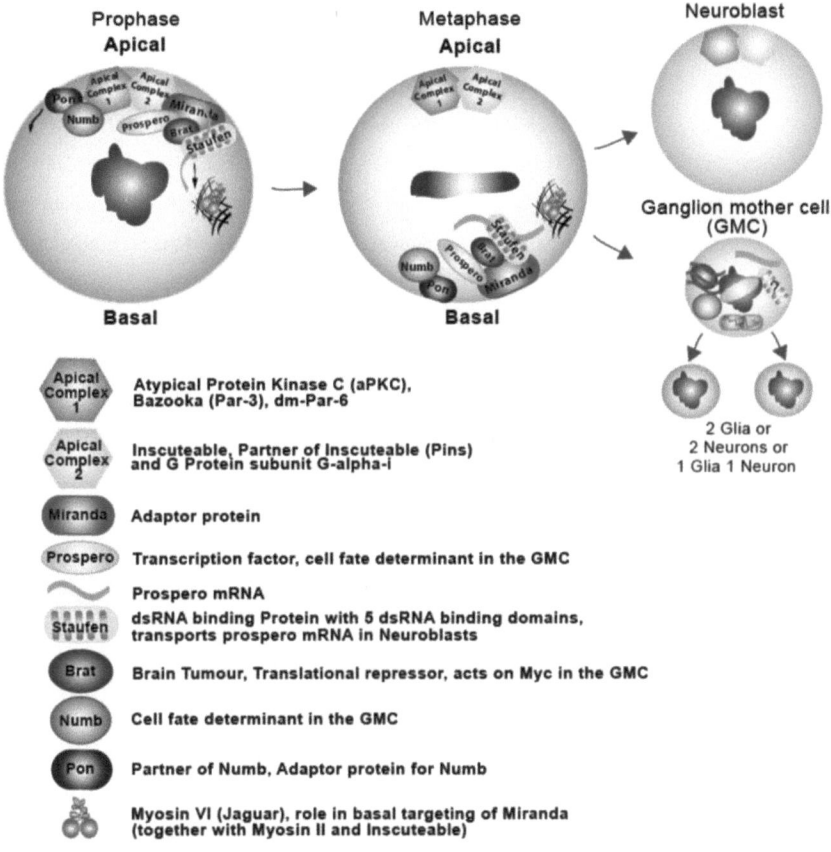

Figure 3. Schematic representation of the factors involved in asymmetric cell divisions in *Drosophila* neuroblasts.
The Miranda and the Pon/ Numb complex are localized to the apical pole in interphase/prophase, are transported to the basal pole in metaphase and are then inherited by the GMC. The apical complexes coordinate the basal localisation of Miranda/ Prospero/ Staufen/*prospero* mRNA/ Brat and Pon/ Numb, as well as the re-orientation of the mitotic spindle along the apical/basal axis (not shown). What drives the asymmetric distribution and basal anchoring of the various players is not exactly known, but it was shown that motor proteins are involved, since Miranda interacts with Myosin II and Myosin VI (Petritsch et al, 2003). After cell division the NB daughter inherits stem cell renewing proteins and RNA whereas the GMC inherits cell fate determinants in form of protein and RNA to induce differentiation.

The first complex consists of the adaptor protein Miranda and its cargo, the GMC transcription factor Prospero (Pros) as well as its RNA, the double –stranded RNA binding protein Staufen and the translational inhibitor Brat (Betschinger et al, 2006; Fuerstenberg et al, 1998; Ikeshima-Kataoka et al, 1997; Lee et al, 2006b; Li et al, 1997; Matsuzaki et al, 1998; Shen et al, 1997). The other complex consists of the cell fate determinant Numb and its adaptor Partner of Numb (Pon) (Lu et al, 1998) (Figure 3).

The asymmetric localization of the cell fate determinants changes throughout the neuroblast cell cycle. At interphase, the Miranda complex accumulates apically while Pon and Numb are uniformly cortical. From prophase onwards, both complexes form a basal crescent. After cytokinesis Miranda and Pon release their cargoes, which can then carry out their assignation in determining the fate of the GMC (Figure 3).

Numb acts as a repressor of the Notch pathway in the GMC (Le Borgne et al, 2005; Schweisguth, 2004). The transcription factor and homeodomain protein Prospero only enters the nucleus in the GMC, although also expressed in the neuroblast. It has been shown that Prospero binds upstream of over 700 genes, many of which are involved in neuroblast self-renewal or cell-cycle control.

Prospero can also induce the expression of neural differentiation genes which indicates its role as a transcriptional activator and inhibitor (Choksi et al, 2006). The localized *pros* RNA in turn, which is transported by Staufen (that binds to Miranda), is not required for the specification of the GMC. The reason for its asymmetric localization seems to reflect a backup mechanism for the Prospero protein supply.

A third cell fate determinant transported by Miranda, namely brat (brain tumour), was more recently identified (Bello et al, 2006; Betschinger et al, 2006; Lee et al, 2006b). Brat was previously shown to act as inhibitor of ribosome biogenesis and cell growth (Frank et al, 2002), and as a posttranscriptional inhibitor of dMyc (Betschinger et al, 2006).

During embryogenesis, Brat cooperates with Pros to specify the GMC fate. While in *pros* mutants only a small subset of GMCs is affected, pros/brat double mutants show an almost complete loss of GMCs (Betschinger et al, 2006). This underscores the importance of these two proteins for GMC specification and indirectly the role of their adaptor protein Miranda, in properly localizing them.

Apart from the *prospero* mRNA, two other mRNAs which code for proteins involved in the asymmetric cell division machinery are known to be asymmetrically localized. These are *inscuteable* and *miranda* mRNA (Hughes et al, 2004; Schuldt et al, 1998).

Like the protein, *inscuteable* mRNA also localizes apically in the neuroblast (Knirr et al, 1997; Li et al, 1997). It could be shown that the *inscuteable* mRNA is posttranscriptionally regulated by Abstrakt (Abs), which is a member of a family of RNA-dependent ATPases called DEAD-box proteins (Irion et al, 2004).

Recently the localization machinery of *inscuteable* could be unravelled. The Egalitarian (Egl)/ Bicaudal-D (BicD/ dynein mRNA transport machinery (Bullock & Ish-Horowicz, 2001) mediates the apical localization of the *inscuteable* mRNA transcripts in neuroblasts. This localization seems to be required for efficient apical targeting of Inscuteable protein (Hughes et al, 2004).

In contrast, the mechanism or the significance of the apical cytoplasmic *miranda* mRNA localization (throughout the cell cycle) remains unclear.

The adaptor protein Miranda was identified in a yeast two-hybrid screen in 1997, where Prospero was used as bait (Shen et al, 1997). In embryos homozygous for a null allele of *inscuteable*, Miranda and its cargo are unable to form crescents at all, or they form crescents that are randomly localized across the cell membrane (Shen et al, 1997). This lead to the conclusion that correct asymmetric localization and crescent formation requires Inscuteable and the Par complex.

Although the Miranda protein itself is not conserved, it inherits several conserved domains (Figure 4). It contains several coiled-coil repeats in its central region allowing the interaction with Staufen, Prospero, Numb and Inscuteable (Fuerstenberg et al, 1998; Shen et al, 1998). The C-terminus contains 7 consensus PKC sites as well as signals for timely degradation and therefore release of its cargo in the GMC. These signals correspond to four potential destruction boxes. A destruction box is a 9 aa motif that is conserved among the N termini of A- and B- type cyclins (King et al, 1996). These destruction boxes are required for the cell-cycle dependent degradation of these cyclins by an ubiquitin dependent pathway in anaphase during mitosis (Yamamoto et al, 1996), whereas Miranda disappears in the GMC after mitosis has completed (Shen et al, 1997).

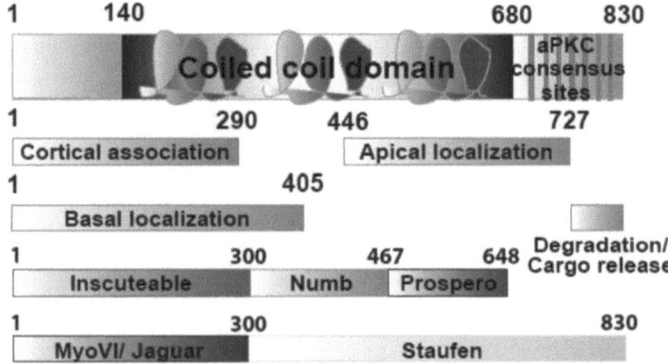

Figure 4. Domain organization and functional assignment of the Miranda protein.
Miranda is an 830 amino acid protein. The central part of the protein (residues 140-680) consists of a coiled-coil domain, which is known to be required for Numb/ Prospero and Staufen interaction, whereas the n-terminal 300 amino acids are needed for Inscuteable/ MyoVI (Jaguar) interaction (Fuerstenberg et al, 1998; Petritsch et al, 2003; Shen et al, 1998). The C-terminus contains seven aPKC consensus sites as well as destruction boxes which are necessary for Miranda degradation and Cargo release in the newborn GMC. The first 300 amino acids were shown to be necessary for cortical localization, whereas basal localization requires a slightly larger portion of the N-terminus (adapted from (Fuerstenberg et al, 1998; Shen et al, 1998).

The dynamic localization pattern of Miranda throughout the cell cycle and its dependence on an intact actin cytoskeleton suggested the involvement of myosin motors for this process. In fact, Miranda complexes containing Myosin II and Myosin VI (Jaguar) could be identified (Petritsch et al, 2003). Petritsch and colleagues could show a co-localization of Jaguar with Miranda on the neuroblast cytoplasm, although no co-migration in a basal crescent could be observed.

Barros and colleagues could show that Miranda is excluded from the apical cortex by the nonmuscle Myosin II, which is restricted to the apical membrane by the tumour suppressor Lgl. During pro- and metaphase this activated Myosin II prevents Miranda from localizing apically (Barros et al, 2003).

A recent publication could specify some aspects of Miranda movement. In this publication we could show that in contrary to former descriptions, Miranda forms an apical crescent in interphase, whereas in prophase Miranda shows a rather ubiquitous cytoplasmic localization (Erben et al, 2008).

Furthermore, it could be clarified that Miranda reaches the basal cortex by passive diffusion throughout the cytoplasm rather that by long-range Jaguar-directed transport. Nevertheless, Jaguar acts downstream of Myosin II to deliver diffusing Miranda to the basal cortex (Erben et al, 2008). The basal anchor protein for Miranda is not yet discovered.

1.2.3. Role of Cell Cycle Regulators in Neuroblast Cell Divisions

1.2.3.1. Cdc2

The first indication that cell cycle regulators might also control certain aspects of the asymmetric cell division of neural progenitors came from a study on Cdc2 (Ashraf & Ip, 2001). Cdc2 (cell division cycle 2) in complex with A-or B type cyclins, is necessary to drive cells from G2 into mitosis.
Tio and colleagues could show that Prospero and Inscuteable exhibited a defective localization in Cdc2 mutant neuroblasts. Analysis of mutants that express attenuated levels of Cdc2 (reduced levels not efficient to prevent the cells from entering into mitosis), showed a failure of asymmetrically localizing the apical (Inscuteable and Bazooka) and the basal components (Miranda and Pon, of which proper localization depends on the correct localization of the apical components) (Schober et al, 1999). As a consequence, asymmetric divisions are converted to symmetric divisions.

1.2.3.2. Aurora A and Polo Kinases

Two highly evolutionary conserved kinases, Aurora A and Polo, have recently been shown to play a role in the asymmetric cell division machinery (Lee et al, 2006a; Wang et al, 2006).
Aurora kinases regulate mitosis and meiosis in all eukaryotes and they are linked to oncogenesis (Meraldi et al, 2004).

Deregulation of Aurora kinase impairs spindle assembly, checkpoint function and cell division, which leads to a missegregation of individual chromosomes and polyploidization. Furthermore Aurora kinases are frequently overexpressed in cancers (Meraldi et al, 2004).

In *Drosophila,* Aurora A is required for centrosome maturation, cell cycle progression, Numb protein localization during sensory organ precursor asymmetric cell division and astral microtubule length in *Drosophila* Schneider cells and larval neuroblasts (Giet et al, 2002).

Aurora A was shown to act as a tumour suppressor by suppressing NB self-renewal and promoting neuronal differentiation during larval brain development (Wang et al, 2006).

Mutant analysis revealed that a loss of Aurora A produces 2 self-renewing daughter cells leading to an excess of neuroblasts at the expense of differentiated neurons. Aurora A is required for the asymmetric localization of aPKC to the apical cortex and promotes Numb basal localization. Although aPKC mutants affect all known basal localized proteins (Miranda, Numb, Brat and Pros), *auroraA* mutants only show a detectable change in Numb localization.

The evolutionary conserved kinase Polo (Cdc5 homolog) is active during mitosis and the mammalian counterparts have been implicated in acting as tumour suppressors (Barr et al, 2004; van de Weerdt & Medema, 2006). In *Drosophila,* the Polo kinase peaks from late anaphase to telophase, later than the peak of Cdc2-cyclin B kinase.

Loss of Polo leads to a mislocalization of the localized proteins aPKC and Numb in *Drosophila* larval neuroblasts, which leads to a neuroblast overgrowth phenotype (Wang et al, 2007). Miranda, Brat and Prospero localization is not affected.

It was shown that Pon is phosphorylated by Polo and that this phosphorylation regulates the asymmetric localization of Pon and therefore also of Numb. The fact that Numb inhibits neuroblast self-renewal by antagonizing Notch signalling in the GMC explains why its mislocalization in *polo* leads to the neuroblast overgrowth phenotype (Wang et al, 2007; Wang et al, 2006).

1.2.3.3. Cyclin E

Cyclin E (CycE) together with CDK2 provides a complex that regulates the G1 to S-phase transition during the cell cycle. The role of Cyclin E in *Drosophila* neuroblasts has been recently studied in detail in the NB6-4 lineage (Berger et al, 2005).
The NB6-4 lineage shows segment specific differences in its progeny outcome. The thoracic NB6-4 (NB6-4t) generates neurons and glia cells, whereas abdominal NB6-4 (NB6-4a) generates only glia cells (Figure 5).

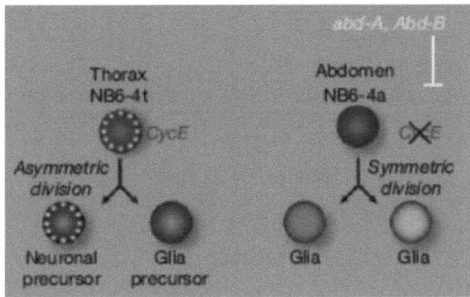

Figure 5 Mechanism of segment specificity of the NB6-4 lineage by differential CycE expression.
In the NB6-4t lineage CycE is distributed only to the neuronal precursor, whereas Pros and *gcm* are asymmetrically distributed to the glia cell (not shown). The expression of CycE in the NB6-4a is inhibited by Abd-A and Abd-B thus leading to a symmetric distribution of Pros and *gmc* to two glia cells (not shown)(adapted from (Berger et al, 2005)).

This difference results from the first asymmetric first division of NB6-4t. In this division, *glial cells missing* (gcm) transcripts are distributed to both daughter cells, whereas in the cell that functions as neuronal precursor, *gcm* is rapidly removed. Prospero is transferred asymmetrically into only one cell, where it is needed to maintain and enhance the expression of *gcm*, thereby promoting the glial cell fate.
The NB6-4 abdominal lineage divides symmetrically. Therefore both daughter cells express *prospero* and *glial cells missing*.

In the NB6-t lineage, *cycE* mRNA is only detected in the neuronal, but not in the glia precursor. In contrast, the NB6-4 abdominal lineage shows no *cycE* expression at all (Figure 5).

The two homeotic genes Abdominal A and Abdominal B, which are expressed in the abdominal segments of the *Drosophila* embryo, specify the NB6-4 abdominal lineage by downregulating levels of CycE. Loss of CycE function causes a homeotic transformation of NB6-4t to NB6-4a. Similar observations where made in other neuroblasts which exhibit segment specific differences. Therefore, in addition to its role in cell proliferation, CycE has a role in specifying the cell fate in certain neuroblast lineages (Berger et al, 2005).

1.2.3.4. The Anaphase Promoting Complex/ Cyclosome

The anaphase promoting complex or cyclosome (APC/C) is a large, multi-subunit complex that promotes the ubiquitination and subsequent proteasome mediated destruction of several proteins during mitosis.

It consists of at least 11 core subunits and functions as an E3 ubiquitin ligase for targeting proteins via the 26S proteasome. APC/C promotes mitotic transitions through several key processes, which include the destruction of mitotic cyclins and inhibitors of chromosome separation as well as the regulation of DNA replication, centrosome dublication and mitotic spindle assembly (reviewed in (Pesin & Orr-Weaver, 2008))

APC/C activity is required for proper asymmetric localization of the adaptor protein Miranda and its associated cargo proteins Staufen, Prospero and Brat in *Drosophila* neuroblasts (Slack et al, 2007).

Mutant embryos of several different subunits of APC/ C showed a mislocalization of Miranda in neuroblasts. Apparently, proper localization of Pon/ Numb or the apical complex members does not require APC/ C activity.

Miranda protein was shown to be ubiquitylated in cultured cells and larval neuroblasts. Nevertheless, only a monoubiquitylation could be demonstrated (only polyubiquitylated proteins are degraded by 26S proteasome). It remains unclear, if that ubiquitylation is APC/C dependent.

1.2.4. Starting and Stopping Neuroblast Divisions

In *Drosophila*, there are two distinct phases of neurogenesis, embryonic and postembryonic, which are segregated by a pause in neuronal proliferation known as quiescence. Once the neuroblasts have generated their embryonic lineages, the first phase of neurogenesis ends. They exit the cell cycle and enter a G0 or G1- like quiescent state (Truman & Bate, 1988). It is not clear what triggers neuroblasts to exit the cell cycle. In the GMC however, Prospero seems to play an important role in preventing them to divide more than once, by repressing cell cycle genes (Choksi et al, 2006; Li & Vaessin, 2000).

The segregation of the neuroblasts takes place during the embryonic stages 8-11 in 5 waves (SI to SV) (see Figure 2 and Figure 6).
Most embryonic neuroblast divisions cease by embryonic stage 14, but in the thorax a few remain mitotically active until stage 16 (Prokop & Technau, 1991). In the brain, one lateral neuroblast and four mushroom-body neuroblasts divide continuously through to the pupal stages, therefore escaping quiescence completely (Ito & Hotta, 1992).

A

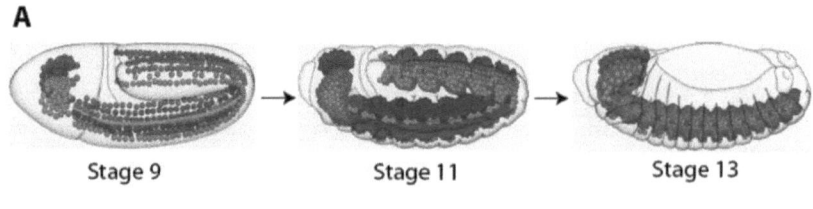

Stage 9   Stage 11   Stage 13

B

Embryonic Stage	Time (after egg laying)	Developmental events
1-4	0:00-2:10 h	Cleavage
5	2:10-2:50 h	Blastoderm
6-7	2:50-3:10 h	Gastrulation
8-11	3:10-7:20 h	Germ band elongation
12-13	7:20-10:20 h	Germ band retraction
14-15	10:20-13:00 h	Head inv./ dorsal clos.
16-17	13:00-22:00 h	Differentiation

Figure 6. Schematic representation of the CNS development at different embryonic stages. Developing CNS is shown in violet and the mesectoderm in blue (A). The embryonic development of *Drosophila* has been subdivided into 17 stages by Volker Hartenstein and Jose Campos-Ortega. These stages are listed in table B. Delamination of embryonic neuroblasts occurs between stages 8-11 and most of them divide until stage 14. Staging according to these authors has become a general reference in *Drosophila* research (Campos-Ortega, 1985).

One important system regulating the number of neuroblast divisions is that the fate of embryonic neuroblasts changes over time. This is triggered by the sequential expression of a defined set of transcription factors (Isshiki et al, 2001). GMCs and their progeny (neurons and glia) maintain the expression profile of the transcription factor which was present at their time of birth providing it with a temporal label (Isshiki et al, 2001).

So far, four members of this transcription factor series have been identified in the early embryos, which are expressed in the following order: Hunchback → Krüppel → Pdm1 → Castor (Figure 7).

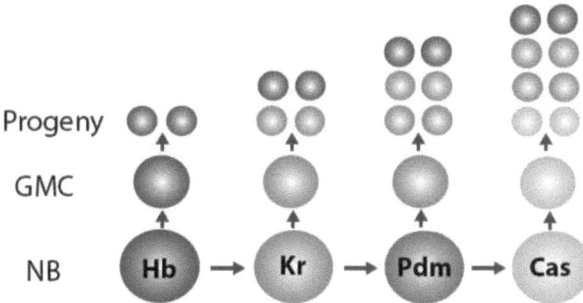

Figure 7. Transcription factor switching during *Drosophila* embryonic neuroblast divisions.
Dividing neuroblasts in the embryonic CNS switch through a series of transcription factors. These are normally expressed in the order: Hunchback → Krüppel → Pdm → Castor. The GMC maintains the same transcription factor as the neuroblast it derived from and forwards it to its progeny after division. Therefore the "oldest" progeny are localised deeper in the embryo and express Hunchback, whereas "younger" progeny are situated more superficial and express Castor.

However, in some neuroblast lineages only subsets of this sequence are expressed. A fifth member of the temporal series is the nuclear receptor Seven-up, which is transiently expressed in early embryonic neuroblasts and was shown to be required for the Hunchback → Krüppel switch (Kanai et al, 2005; Mettler et al, 2006).

The transitions from one transcription factor to another depend on cell cycle progression and are stabilized by negative cross-regulatory interactions (Isshiki et al, 2001; Kambadur et al, 1998). Studies in cultured neuroblasts suggest that the expression of another transcription factor, Grainyhead (Grh), follows Castor (Brody & Odenwald, 2000; Uv et al, 1997). Experiments to keep a neuroblast "youthful", by persistently expressing the *hunchback* or the *krüppel* gene, produced many more cells than normal in the neuroblast lineage (Isshiki et al, 2001).

These findings suggested that embryonic neuroblasts might exit the cell cycle and become quiescent only when they have finished switching through their normal temporal sequence of factors.

Programmed cell death provides one irreversible mechanism for ensuring that embryonic neuroblasts stop dividing. At the end of embryogenesis neuroblast apoptosis is mainly found in the abdomen, which provides a large step towards forming the adult CNS.

This leads to a survival of only 3 out of 30 initial abdominal neuroblasts and about 20 out of 30 thoracic neuroblasts per hemisegment when the larva hatches (Prokop et al, 1998; White et al, 1994).

Following the quiescent period, most postembryonic neuroblasts resume asymmetric divisions, expressing many of the asymmetric cell fate determinants in a similar pattern as in the embryo (Ceron et al, 2001).

1.3. RNA Localization

1.3.1. General Role of RNA Localization

In *Drosophila melanogaster*, 71% out of 3370 genes, encode subcellularly localized mRNAs (Lecuyer et al, 2007). This amazing finding might lead to the conclusion that mRNA localization is a major mechanism for controlling cellular architecture and function.

It is economically advantageous for the cell to localize mRNAs instead of their corresponding protein. In fact, each localized mRNA can facilitate many rounds of protein synthesis, thereby avoiding the significant energy costs of moving each protein molecule individually (Jansen, 2001).

Localized mRNAs can serve several biological functions, like the establishment of morphogen gradients (Driever & Nusslein-Volhard, 1988; Ephrussi et al, 1991; Gavis & Lehmann, 1992), the segregation of cell-fate determinants (Broadus et al, 1998; Gore et al, 2005; Hughes et al, 2004; Li et al, 1997; Long et al, 1997; Melton, 1987; Neuman-Silberberg & Schupbach, 1993; Simmonds et al, 2001; Takizawa et al, 1997; Zhang et al, 1998), or the targeting of protein synthesis to specialized organelles or cellular domains (Adereth et al, 2005; Lambert & Nagy, 2002; Lawrence & Singer, 1986; Mingle et al, 2005; Zhang et al, 2001).

Mislocalized mRNA lead to mislocalized proteins which can have severe consequences. This could be demonstrated for the case of *nanos* and *oskar*, two localizing mRNAs in the *Drosophila* oocyte. A mislocalization of either one would induce the development of a second abdomen in the place of head and thorax (Ephrussi et al, 1991; Gavis & Lehmann, 1992).

1.3.2. Mechanisms of mRNA Localization

Most of the characterized examples of mRNA localization occur by active transport through the cytoskeleton. To be localized, an mRNA must contain *cis*-acting localization elements that are recognized by specific RNA binding proteins, which couple the mRNA to the localization machinery.

The active transport of the mRNA via a motor can take place along microfilaments or microtubules, depending on the type of motor protein like dynein, myosin or kinesin.

Apart from active transport, mRNAs can also be localized by an indirect mechanism, which is degrading all the transcripts that are not at the correct place. This could be for example shown for the *hsp83* mRNA, which thereby is restricted to the pole plasm of the posterior *Drosophila* oocyte (Ding et al, 1993). This is accomplished by two cis acting elements in the 3′ UTR of the RNA, namely a degradation element that targets the mRNA for destruction in all regions and a protection element that stabilizes the mRNA at the posterior (Bashirullah et al, 1999).

Another mechanism of localizing RNAs is passive diffusion through the cytoplasm and capturing by a localized anchor. This is known to be the case for several transcripts, like *nanos*, *gcl* (germ cell-less) and *cyclin B* mRNAs, which become enriched in the *Drosophila* pole plasm (Jongens et al, 1992; Raff et al, 1990; Wang et al, 1994).

After all, the most obvious way to localize an mRNA is by just synthesizing it locally. Nevertheless, this seems to be a rare mechanism. It contributes to one aspect of *gurken* mRNA localization on one side of the *Drosophila* oocyte nucleus, although other mechanisms for the proper localization of this mRNA exist, including dynein-dependent movement (MacDougall et al, 2003; Saunders & Cohen, 1999; Thio et al, 2000).

1.3.3. Staufen has a Conserved Role in RNA Localization

The *staufen* gene was identified initially as a maternal factor required for the correct formation of the anteroposterior axis in *Drosophila* (St Johnston et al, 1991; St Johnston et al, 1989).

The Staufen protein contains double-stranded RNA-binding domains (dsRBDs) that were shown to bind individually but non-specifically to dsRNA (St Johnston et al, 1992). The Staufen protein is responsible for localizing RNAs in different cell types in *Drosophila* (see next chapter). Homologues of Staufen could be identified in the clawfrog, rat, mouse, and human (Kiebler et al, 1999; Marion et al, 1999; Wickham et al, 1999) (Figure 8).

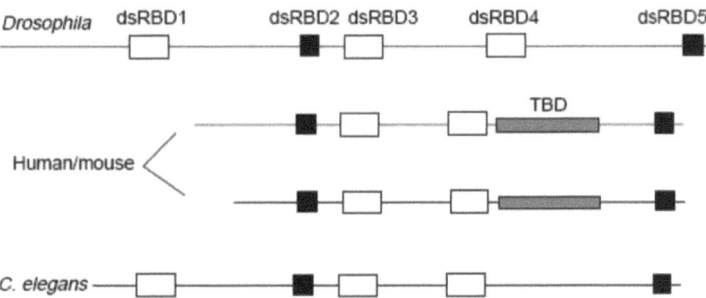

Figure 8. Domain structures of Staufen homologues.
The *Drosophila* gene contains two short (black) and 3 long (white) domains which can bind to dsRNA. The human /mouse counterparts lack the first dsRBD and contain an accessory tubulin binding domain (TBD, grey). The *C.elegans* gene contains 5 dsRBDs and lacks the TBD, which resembles the structure of the *Drosophila* Staufen (Roegiers & Jan, 2000).

Xenopus laevis Staufen protein localizes to the vegetal pole with *Vg1* and it could be shown that this movement depends on kinesin (Yoon & Mowry, 2004). *C. elegans* also contains an uncharacterized open reading frame that shows similarities to human and mammalian Staufen (Wickham et al, 1999).

Drosophila Staufen consists of 5 dsRBD, whereas mammalian Staufen lacks the first dsRBD (Marion et al, 1999; Wickham et al, 1999).

Mammalian Staufen contains a putative microtubule-associated protein (MAP-1B)-binding domain (MBD) that is missing in the *C. elegans* and *Drosophila* proteins.

1.3.4. Staufen Dependent RNA Localization in *Drosophila*

The *Drosophila* oocyte is a well characterized example showing the importance of localized transcripts for the determination of the anterior posterior axis.

The anterior determinant is encoded by *bicoid* mRNA that localizes to the anterior pole and is necessary to define where the head and the thorax develop (Berleth et al, 1988). The RNA is translated after fertilization to produce a homeodomain transcription factor, which diffuses posteriorly to form a morphogen gradient. This gradient patterns the anterior of the embryo, first by activating the transcription of various zygotic gap genes and second by directly binding to *caudal* mRNA and repressing its translation (Rivera-Pomar et al, 1996).

Interestingly, while the anterior cytoplasm contains only one maternal mRNA determinant, the pole plasm at the posterior of the egg contains several mRNAs that play essential roles in the determination of the abdomen and the pole cells which represent the founders of the germline lineage (D., 1993).

The first mRNA that reaches the posterior of the oocyte is *oskar*, which in contrast to *bicoid* is directly translated when it reached its destination. The Oskar protein nucleates the assembly of the pole plasm, which contains the abdominal determinant *nanos* mRNA (Wang & Lehmann, 1991). The Nanos protein is translated after fertilization and forms a gradient that directs the formation of the abdomen by repressing the translation of the *hunchback* mRNA (Hulskamp et al, 1989; Struhl, 1989).

The localization of *oskar* and *bicoid* mRNAs was found to depend on maternally provided Staufen. The protein was shown to co-localize with *oskar* mRNA at the posterior pole of the *Drosophila* oocyte and with *bicoid* mRNA at the anterior of the egg and the early embryo (Ferrandon et al, 1994; St Johnston et al, 1991).

The *oskar* localization depends on the polarized microtubule cytoskeleton and the plus-end directed microtubule motor kinesin, suggesting that Staufen may play a role in coupling the mRNA to kinesin, which then transports the complex along microtubules (Brendza et al, 2000).

In the *Drosophila* neuroblast, Staufen is required for the asymmetric localization of prospero mRNA. At metaphase, Staufen/ *prospero* are localized from the apical to the basal cortex by the adaptor protein Miranda. In contrast to other examples of Staufen dependent mRNA localization, this movement depends on the actin cytoskeleton and not on microtubules (Broadus et al, 1998; Li et al, 1997; Matsuzaki et al, 1998; Schuldt et al, 1998).

Miranda also binds to and transports the Prospero protein as well as the translational repressor Brat (Betschinger et al, 2006; Ikeshima-Kataoka et al, 1997; Lee et al, 2006b). The complex is then inherited by the GMC, where Miranda is degraded and the cell fate determinants are released to specify the cell (Betschinger et al, 2006; Hirata et al, 1995; Knoblich et al, 1995; Lee et al, 2006b).

Prospero needs to be transported to the GMC, because the cell cannot transcribe this gene itself (Broadus et al, 1998). The homeodomain transcription factor and cell fate determinant Prospero represents a clear example of redundancy between protein and RNA localization. It could be shown that in *staufen* mutants the GMC can still develop normally (Broadus et al, 1998).

Therefore, the accessory *prospero* RNA serves as an assurance to reinforce the protein localization in the GMC.

1.4. Goals of the Thesis

During my thesis, I was focusing on 2 objectives.

At first, I wanted to identify and characterize the possible basal anchor protein and further proteins, involved in the asymmetric localization of the Miranda protein complex in *Drosophila* neuroblasts. To do that, I followed two strategies. These were GST-pulldown experiments using the N-terminal Miranda domain as bait and immunoprecipitations of Miranda complexes. Both were carried out on *Drosophila* embryo extracts and co –purified proteins were identified by mass spectrometry.

The second objective of this work was to identify and characterize further RNAs that might be transported by Miranda/ Staufen to the GMC. This was carried out by immunoprecipitating the Miranda protein complex in a first step, then eluting the bound RNAs and reverse transcribing them. A set of candidate primer pairs was then tested on the obtained cDNA pool. The positively tested candidates where further examined by whole mount *in situ* hybridization and by immunofluorescence.

2. Materials and Methods

2.1. Materials

2.1.1. Chemicals

Biozym Scientific GmbH, Hess. Oldendorf, Germany;
Carl Roth GmbH Co., Karlsruhe, Germany;
Fluka Chemie GmbH, Buchs, Germany;
Merck KGaA, Darmstadt, Germany;
Sigma-Aldrich Chemie GmbH, Steinheim, Germany;

All the buffers used for experiments under RNAse free conditions were prepared with DEPC treated water from Roth and sterile filtered. All other buffers were prepared with distilled and autoclaved H_2O and sterile filtered after preparation.

2.1.2. Enzymes

New England Biolabs (NEB) GmbH, Frankfurt am Main, Germany;
Roche Diagnostics GmbH, Mannheim, Germany;
Fermentas GmbH, St. Leon-rot, Germany;

2.1.3. Kits

QIAquick PCR Purification Kit, Qiagen GmbH, Hilden, Germany;
QIAquick Gel Extraction Kit, Qiagen GmbH, Hilden, Germany;
QIAGEN Plasmid Midi Kit, Qiagen GmbH, Hilden, Germany;
QIAprep Miniprep Kit, Qiagen GmbH, Hilden, Germany;
DIG RNA Labeling Kit (SP6/ T7), Roche Diagnostics GmbH, Mannheim, Germany;

SuperScript™ II Reverse Transcriptase Kit, Invitrogen GmbH, Karlsruhe, Germany; ECL Western Blotting Detection System; Amersham Biosciences, Buckinghamshire, UK

2.1.4. Antibodies

2.1.4.1. Commercially available Antibodies

- Mouse α BrdU, Roche Diagnostics GmbH
- Mouse α Digoxigenin, Roche Diagnostics GmbH
- Sheep α Digoxigenin, Roche Diagnostics GmbH
- Mouse α GFP (B-2): sc-9996, Santa Cruz Biotechnology Inc., California USA
- Rabbit α-phospho-Histone H3 (Ser10), Upstate, California USA
- Rabbit α cleaved Caspase-3 (Asp175), Cell Signaling
- TOTO®-3 iodide, Molecular Probes (Invitrogen), Karlsruhe, Germany
- Normal Goat Serum (NGS), Jackson ImmunoResearch, West Grove, USA

Unless otherwise noted, mouse primary antibodies were purchased from the Developmental Studies Hybridoma Bank (DSHB), Iowa City, USA. The fluorescently coupled secondary antibodies for immunostainings, as well as the horseradish peroxidase-conjugated secondary antibodies for westernblots were purchased from Jackson ImmunoResearch, West Grove, USA.

2.1.4.2. Non-commercial Antibodies

- Mouse α Myosin VI (3C7); Kathryn G. Miller, Washington University, St. Louis
- Rabbit α Miranda ((Shen et al, 1997) raised against the N-terminal peptide sequence: SLPQRLRFRPTPSHTDTATGSGS)
- Rabbit α Tudor-SN; Gregory Hannon, Cold Spring Harbor Laboratory, Cold Spring Harbor

- Mouse α dFMR1 (6A15); Gideon Dreyfuss, University of Pennsylvania School of Medicine, Philadelphia
- Rabbit α Staufen; Daniel St. Johnston, Wellcome Trust/ Cancer Research UK Gordon Institute, Cambridge UK
- Mouse α Dacapo; Iswar Hariharan, University of Berkeley, California

2.1.5. Fly Stocks

The commercially available flystocks were purchased from the Bloomington *Drosophila* Stock Center, Indiana University, Bloomington Indiana, USA.

2.2. Methods

2.2.1. Standard Laboratory Methods for *Drosophila melanogaster*

2.2.1.1. Laboratory Culture

Standard methods for the laboratory culture of *D. melanogaster* were applied, as described in detail in (Ashburner, 1989).

2.2.1.2. Embryo Collection

To collect embryos for Immunostainings and *in situ* hybridizations, flies were cultured in collection Vials (15 cm in height and 7 cm in diameter). The eggs were collected on apple agar plates. Large quantities of embryos for biochemical experiments (more than 1 g) were collected at the *Drosophila* collection facility of Prof. Peter Becker, Adolf-Butenandt Institute in Munich.

After collection, the embryos were matured at 25°C to the correct stage. They were then collected in a sieve and bleached in 6% sodium hypochlorite for 2-3 min, to remove the chorion (eggshell). After several washing steps, the embryos could be further processed for immunostainings or biochemical experiments.

2.2.2. Methods in Molecular Biology

2.2.2.1. Oligonucleotides

Name	Sequence (5´-3´)	Application
Dacapo ISH F	CGTGACCTCTTCGGTAGCTC	DIG-labeled RNA probes for ISH
Dacapo ISH R	GGATCCTAATACGACTCACTATAGGGAGAACTCGTCCTGGGAGTCGTAA	DIG-labeled RNA probes for ISH
Inscuteable ISH F1	TAGCATGAAGCTGGACGATG	DIG-labeled RNA probes for ISH
Inscuteable ISH R1	GGATCCTAATACGACTCACTATAGGGAGAGACCCACTTTTTGGTGTGCT	DIG-labeled RNA probes for ISH
Prospero ISH F	GGATCCTAATACGACTCACTATAGGGAGACTGATCGACGATTCGGAAAT	DIG-labeled RNA probes for ISH
Prospero ISH R	CATACGATTTAGGTGACACTATAGAAGAGGTTGTTGCTGTTGCTGCTGT	DIG-labeled RNA probes for ISH
Dacapo (CG1772) F	GTCAGCTTCCAGGAGTCGAG	Candidate PCR ◊; ∆
Dacapo (CG1772)	ACGAACTGGGAGAACTGGTG	Candidate PCR ◊; ∆
Glial cells missing (CG12245) F	GTGATGAGGCCAGGAAACAT	Candidate PCR ∆
Glial cells Missing (CG12245) R	AACATTACGGCCAAACTTCG	Candidate PCR ∆
Hunchback (CG9786) F	CAAAACAGCCTGCAGCATTA	Candidate PCR
Hunchback R	TTCGACATCTGATCGTGCTC	Candidate PCR

Castor (CG2102) F	CCTCAGCTTTTGGAATGCTC	Candidate PCR
Castor (CG2102) R	CATGCTGATCTCGGTCTGAA	Candidate PCR
Krueppel (CG3340) F	CCTCAGCTTTTGGAATGCTC	Candidate PCR
Krueppel (CG3340) R	CATGCTGATCTCGGTCTGAA	Candidate PCR
Hnr27C (CG10377) F	GCAGGTCGAAATCAAGAAGG	Candidate PCR Δ
Hnr27C (CG10377) R	TACATGTCGTATCCGCCGTA	Candidate PCR Δ
Mushroom-body expressed (CG7437) F	GCTTTCACCATCCAAGGAAA	Candidate PCR Δ
Mushroom-body expressed (CG7437) R	TGGGCGCAATGTAGTGATAA	Candidate PCR Δ
Split ends F	GTGCGACTAAACCGTTGGAT	Candidate PCR Δ
Split ends R	TGCACTGCCTTGTTGAGTTC	Candidate PCR Δ
Failed axon connections (CG4609) F	GATCACCTTTGGTCGCAAGT	Candidate PCR Δ
Failed axon connections (CG4609) R	CACTTGTCCTTCATGCGAGA	Candidate PCR Δ
Blistery (CG9379) F	CATCCACATCGACATCCAAG	Candidate PCR Δ
Blistery (CG9379) R	ACTACGGGCAAATTGTACGG	Candidate PCR Δ
Mab-2 (CG4746) F	CGAGGGATTCGACATGCTAT	Candidate PCR Δ
Mab-2 (CG4746) R	CGACTTGCCCTTGAACAGAT	Candidate PCR Δ
Dappled (CG1624) F	CGGAAATTGGTGTGTCAGTG	Candidate PCR Δ
Dappled (CG1624)R	CTCTGCCCATTGGTCAACTT	Candidate PCR Δ
Hs2st (CG10234) F	CTTTCTACGCTTTGGCGAAC	Candidate PCR Δ

Hs2st (CG10234) R	GTCACCCGCAGATGAGATTT	Candidate PCR ∆
ImpL3 (ecdysone inducible gene L3) F	AGAACATCATCCCCAAGCTG	Candidate PCR ∆
ImpL3 (ecdysone inducible gene L3) R	GCAGCTCGTTCCACTTCTCT	Candidate PCR ∆
CG5358 F	TTCGGACGAATCGCTCTACT	Candidate PCR ∆
CG5358 R	AGAAGGCAGCGAACCTGATA	Candidate PCR ∆
CG5235 F	ATTCGTGCTCTACGCCAGTT	Candidate PCR ∆
CG5235 R	AGCTTGATGTGAGTGCAACG	Candidate PCR ∆
Schizo (CG10577) F	TTCCGGATCGGATTCTAGTG	Candidate PCR ∆
Schizo (CG10577)R	TTGAGTCCAACGCGATACTG	Candidate PCR ∆
CG13920 F	CCAATACGATCGTGCTGAAG	Candidate PCR ∆
CG13920 R	CAGCATGAAGGTGAAGACCA	Candidate PCR ∆
Neuralized (CG11988) F	TGGTGAGAAGCTGATTGTGC	Candidate PCR ∆
Neuralized R	CTGGCATTCACATTGACCTG	Candidate PCR ∆
dFMR (CG6203) F	AAGAAGCCCAGAAGGATGGT	Candidate PCR
dFMR (CG6203) R	TTCTCCTCCAGCTCGATGTT	Candidate PCR
Dicer 1 (CG4792) F	TGATCCCGATCTCAAGTTCC	Candidate PCR
Dicer 1 (CG4792) R	TAACTCGGAGCGACGAGAAT	Candidate PCR
Argonaute 1 (CG6671) F	CGAAAGGTGAACCGTGAGAT	Candidate PCR
Argonaute 1 (CG6671) R	TGAGCATCATCTTCCACTGC	Candidate PCR
CDC2 F	GTATAAATGCGCACCGGAAT	Candidate PCR ◊

CDC2 R	TGTGGCAATGAAGAAAATCC	Candidate PCR ◊
Wingless F	CCCAGTTAGTCCGAATGCAG	Candidate PCR
Wingless R	ACAGCACATCAGCCCACAG	Candidate PCR
Notch F	CATGTCCCACGAACTGGAG	Candidate PCR
Notch R	CACTCAGACCGCCCATTC	Candidate PCR
GAPDH F	AATTTTTCGCCCGAGTTTTC	Candidate PCR
GAPDH R	TGGACTCCACGATGTATTCG	Candidate PCR
Drumstick (CG10016) F	GCTGTAATGCGAATCGACAA	Candidate PCR Δ
Drumstick (CG10016) R	AGATTGTCCCGCTGCTTAAA	Candidate PCR Δ
Nerfin 1 (CG13906)F	GAGCCCATTGAAAAGCTCAG	Candidate PCR Δ
Nerfin 1 (CG13906)R	TCAATTTACGCTTCCCTGCT	Candidate PCR Δ
CG7372 F	GGACGCAAAGAGCGTAAGTC	Candidate PCR Δ
CG7372 R	CGCATCTTTAGACGGAAAGC	Candidate PCR Δ
Myb (CG9045) F	GTCCAAGTCCGAGGATGTGT	Candidate PCR Δ
Myb (CG9045) R	AGCTCCAAGTGAGCCTGGTA	Candidate PCR Δ
Dref (CG5853) F	TGTCATCAAGCACGAGGAAG	Candidate PCR Δ
Dref (CG5853) R	CACGGTGGCATACAGCATAC	Candidate PCR Δ
Mcm7 (CG4978) F	GGAGTCTGCTGCATTGATGA	Candidate PCR Δ
Mcm7 (CG4978) R	GCTCATCCGGAATAGTTGGA	Candidate PCR Δ

CDC45L (CG3658) F	CTTTGGAGCTGGAGCAAATC	Candidate PCR Δ
CDC45L (CG3658) R	AGCCGTAGCTCAGCGTAAAG	Candidate PCR Δ
Set F	CAACTTTTGGGTGACCTCGT	Candidate PCR Δ
Set R	AGTTCTGCGATCTCGTCGTT	Candidate PCR Δ
Adar (CG12598) F	GATATCCGTGGAGGTCGATG	Candidate PCR Δ
Adar (CG12598) R	GTTCAAGCGAGGTAGGGTTG	Candidate PCR Δ
Elav (CG4262) F	GTGAAGCTGATACGCGACAA	Candidate PCR Δ
Elav (CG4262) R	AGGCAATGATAGCCCTTGTG	Candidate PCR Δ
Miranda (CG12249) F	GCCTTCTTCATGTCCACCAT	Candidate PCR ■
Miranda (CG12249) R	CCAGCTGACTTTGACCAACA	Candidate PCR ■
Lgl (CG2671) F	GCAATACGCTGCAGTTCAGA	Candidate PCR ■
Lgl (CG2671) R	GCTTACCGCTAACGAAGGTG	Candidate PCR ■
Prospero (CG17228) F	CATGCAGCTGTCCTCCAGT	Candidate PCR ■
Prospero (CG17228) R	AGAGTGCAAAGGAGTCAAGGATT	Candidate PCR ■
Crumbs (CG6383) F	GGAGTACACTGGTGAACTGTGC	Candidate PCR ■
Crumbs (CG6383) R	TGATTCTGGACACATACCATC	Candidate PCR ■
Bazooka (CG5055) F	TCCTCTCAGCAGTCTCACCA	Candidate PCR ■
Bazooka (CG5055) R	CTCAGAGATGCTGCGTCGT	Candidate PCR ■
Gαl (CG10060) F	CGAAGACGAACTTCACGTTG	Candidate PCR ■
Gαl (CG10060) R	CTAGTATTGGCCGAGGACGA	Candidate PCR ■

MATERIALS AND METHODS

Par-6 (CG5884) F	GGAACTCAACTGCCGTGTTT	Candidate PCR∎
Par-6 (CG5884) R	GCCGAAGTTATCGTCGTTGT	Candidate PCR∎
Pon (CG3346) F	ATCATCAGCAGCAGCAACA	Candidate PCR∎
Pon (CG3346) R	ACACCCGAGGGATTGCAG	Candidate PCR∎
Numb (CG3779) F	TTTAGGCGTCGCAAGGAT	Candidate PCR∎
Numb (CG3779) R	GAAGCCGCGTTCGTGATT	Candidate PCR∎
Inscuteable (CG11312) F	GGCGGTTTCTATTCGAGCTT	Candidate PCR∎
Inscuteable (CG11312) R	GGCGAGTAGAACGACGAGTT	Candidate PCR∎
Staufen (CG5753) F	GTTGCTACCATGGGCACTTT	Candidate PCR∎
Staufen (CG5753) R	ACATGGACGATGCGGATAAT	Candidate PCR∎
Pins (CG5692) F	ATGAGCGGGCCCTAAAGTAT	Candidate PCR∎
Pins (CG5692) R	CCTGTGCTCGTAGCTTTTCC	Candidate PCR∎
Dlg (CG1725) F	AAGGGACTGGGCTTCTCAAT	Candidate PCR∎
Dlg (CG1725) R	ATGCACCTGACTTTGGCTCT	Candidate PCR∎
aPKC (CG10261) F	TTTACCTTCGCAACACAATGA	Candidate PCR∎
aPKC (CG10261) R	GGGAGCTGGTGGATCAGTTA	Candidate PCR∎

Δ Candidates chosen from (Brody et al, 2002)

◊ Primers did not work well

∎ Primers designed by Dr. Birgit Czermin

Table 1. Oligonucleotides used for candidate PCR analysis and generation of Digoxigenin labeled RNA probes for *in situ* hybridizations.

2.2.2.2. Preparation of Digoxigenin (DIG)-labeled RNA Probes

PCRs with the in table 1 listed primers for *inscuteable*, *prospero* and *dacapo* were performed with total *Drosophila* cDNA as template.

PCR Reaction (100 µl)

Pros/ Insc	Dacapo	
2.µl	1 µl	Template (*Drosophila* total cDNA)
10 µl	8 µl	25 mM $MgCl_2$
10 µl	10 µl	10×Taq Buffer - $MgCl_2$ + KCl
5 ul	2 µl	10 mM dNTP Mix
5 ul	4 ul	Primer F (10 pmol/ µl)
5 µl	4 µl	Primer R (10 pmol/µl)
1 µl	1 µl	Taq Polymerase
ad 100 µl		H_2O, RNAse free

PCR Conditions

Prospero/ Inscuteable			Dacapo		
95°C	5 min.		95°C	5 min.	
95°C	1 min.		95°C	30 sec.	
60°C	1 min.	**10×**	60°C	30 sec	**30×**
72°C	3 min.		72°C	45 sec.	
95°C	1 min.		72°C	5 min.	
68°C	1 min	**20×**			
72°C	3 min				

The PCR reactions were subsequently purified with the QIAquick PCR purification Kit, followed by Phenol/ Chloroform extraction. The DNA Pellet was dissolved in H_2O_{DEPC}.

RNA labeling Reaction (adapted from the Roche DIG RNA Labeling Kit (SP6/ T7) Protocol)

500 ng	purified PCR product template
2 µl	10× NTP labeling mixture
2 µl	10× Transcription buffer
1 µl	Protector RNase inhibitor
2 µl	RNA Polymerase (SP6 or T7)
ad 20 µl	H_2O_{DEPC}

The labeling reactions were incubated for 2h at 37°C, followed by a DNAse digestion for 15 min at 37°C. The DNAse activity was inhibited by adding 2 µl 0.2 M EDTA (pH 8.0).
The RNA probes were precipitated with 2.5 µl 4 M $LiCl_2$ and 75 µl 100% Ethanol and the obtained RNA Pellet was dissolved in H_2O_{DEPC}. 1/ 10 v/v of DIG-labeled RNA probe was heated in RNA loading dye, followed by addition of 1 µl 1:1000 diluted SYBR gold solution (Molecular probes). The reactions were electrophoresed on a native 1% agarose gel, prepared with 1× TBE under RNAse free conditions. The residual dissolved probes were aliquoted and frozen at -80°C.

2.2.2.3. RNA Preparation and Reverse Transcription

Following Miranda immunoprecipitation, the protein A-sepharose beads (Amersham) were incubated for 10 min in 150 µl Trizol (Invitrogen) at RT. After the incubation, 50 µl chloroform were added and the reactions were centrifuged for 30 min at 4°C. The upper phase containing the RNA was precipitated with isopropanol and glycogen (Roche) as carrier at -20°C O/N. The RNA pellets were dissolved in 20 µl nuclease free H_2O (Fermentas). Residual template DNA was removed by DNAse digestion.

Reverse Transcription Reaction (SuperScript II Kit, Invitrogen)

1. 10 µl IgG/ IP dissolved RNA + 1 µl (0.5 µg) Oligo d (T) 18 mRNA Primer (NEB)
2. 10 µl Input dissolved RNA + 10 µl nuclease free H_2O (Fermentas) + 3 µl (1.5 µg) Oligo d(T) 18 mRNA Primer (NEB)

The reactions were incubated for 10 min. at 70°C and then cooled down on ice. Then the following components were added:

IgG/ IP RNA	Input RNA	
4 µl	8 µl	5× FS Buffer
2 µl	4 µl	0.1 M DTT
2 µl	4 µl	10 mM dNTP Mix
1 µl	2 µl	SuperScript II Enzyme

The reactions were incubated for 2 h at 42°C, followed by a heat inactivation of the enzyme for 15 min. at 70°C.

2.2.2.4. Candidate PCR Analyses

PCR Reaction (25 µl)

1 µl	Template (RT-PCR reaction, not diluted, 1:10, 1:100 or 1:1000)
2 µl	25 mM $MgCl_2$
2.5 µl	10× Taq Buffer + KCl – $MgCl_2$
0.5 µl	10 mM dNTP Mix
1 µl	Primer F (10 pmol/ µl)
1 µl	Primer R (10 pmol/ µl)
0.25 µl	Taq Polymerase
ad 25 µl	H_2O, nuclease free

PCR Conditions

95°C	5 min.	
95°C	30 sec.	
60°C	30 sec.	**40×**
72°C	30 sec.	
72°C	5 min.	

2.2.3. Methods in Biochemistry

2.2.3.1. SYPRO Ruby Protein Staining

After the proteins were separated by SDS-PAGE, The gel was incubated for 30 min. in fixing solution (10% methanol, 7% acetic acid, 83% H_2O), before the staining in SYPRO Ruby solution (BioRad) O/ N. The gel was then rinsed in fixation solution for 30 min, followed by a 30 min wash step in H_2O before imaging.

2.2.3.2. Preparation of GST-Miranda Beads

2.2.3.2.1. Protein Expression

The glycerol stocks BL21(DE3)/ GST-Miranda$_{N298}$ (N-terminal Miranda protein domain, required for asymmetric localization and cortical association fused to GST) and BL21 (DE3)/ GST (pGEX-4T-1 vector) were streaked on LB-ampicillin plates.
The next day, one colony of each strain was used to inoculate 5 ml LB-Medium containing 100 µg/ ml ampicillin. The preculture was incubated O/N at 37°C with vigorous shaking.

The next morning 250 ml LB- medium supplemented with 100 µg/ ml ampicillin was inoculated with 2.5 ml preculture and incubated at 37°C on a shaker until OD600 reached 0.6 to 0.8 (2.5 to 3 h). The protein expression was induced with 0.5 mM IPTG for 4.5 h at 37°C. The cultures were then centrifuged for 15 min at 5000 rpm, the supernatant was discarded and the pellet frozen at -20°C.

2.2.3.2.2. Protein Purification and Preparation of Beads

The bacteria pellets were thawed on ice and resuspended in 10 ml pre-chilled 1× PBS. The cells were sonified on ice to prevent protein degradation (Pulse 70, Output 10-15, 6× 30 sec). After sonification, Triton-X 100 was added to a final concentration of 1%. The cell extract was incubated on a nutator at 4°C for 30 min. To remove the debris, the extract was centrifuged and the supernatant (SN) was decanted into a new vial. Meanwhile 1 ml glutathione sepharose beads slurry was equilibrated with Lysis Buffer (1× PBS, 1% Triton-X 100). The SN was incubated with the equilibrated beads for 1 h on a rotating wheel at 4°C. The beads were washed 3 times for 15 min each with pre-cooled Washing Buffer (1× PBS, 0.1% Triton-X 100) and then stored in the same buffer with included protease inhibitor (Roche) at 4°C.

2.2.3.3. GST Pull-Down Experiments

2.2.3.3.1. Preparation of *Drosophila* Embryo Extract

***Drosophila* Extraction Buffer (DXB)**

25 mM	HEPES pH 6.8
50 mM	KCl
1 mM	$MgCl_2$/ DTT
250 mM	Sucrose
1×	Protease inhibitor (Roche)

Large quantities of *Drosophila* embryos were collected (5-10 g per collection). After bleaching, the embryos were homogenized 1:2 with freshly prepared DXB in a dounce homogenizer. After 10 strokes with the loose and 10 strokes with the tight pestle, Triton-X100 was added to a final concentration of 0.5% and the homogenate was incubated for 1 h on a nutator at 4°C. The homogenate was then centrifuged for 10 min. at 1500 rpm and 4°C, filtered through a Schüll paper filter and then centrifuged again. The SN from the last centrifugation step was subjected to GST pull-down experiments.

2.2.3.3.2. GST-Pull-Down

The SN from step 2.2.3.3.1 was split 1:1 and incubated with either 200 µl equilibrated GST-Miranda$_{N298}$ coated glutathione sepharose beads, or 200 µl equilibrated GST-coated beads (as control) for 4 h at 4°C on a rotating wheel. After the incubation, the beads were washed twice with DXB and once with DXB containing 100 mM KCl. Bound proteins were eluted with 3 ml DXB containing 1M KCl for 10 min at room temperature.

The eluate was adjusted to 10ml with DXB to diminish salt concentration. 1 ml 0.15% deoxycholic acid (DOC) was added to the diluted eluate and the reaction was incubated for 10 min. at RT. Then, 1 ml of 72% trichloroacetic acid (TCA) was added followed by an incubation O/ N at RT. The next day, the precipitated proteins were centrifuged for 1 h at 4500 rpm and 4°C. The SN was discarded and the pellet was resuspended in 1× SDS loading dye. The samples were heated for 5 min at 95°C and analysed by SDS-PAGE.

2.2.3.4. Immunoprecipitation Experiments

2.2.3.4.1. Preparation of *Drosophila* Embryo Extract

Large quantities of embryos were collected *Drosophila* (5-10 g per collection) and frozen in liquid nitrogen after bleaching. The frozen embryos were ground to a fine powder with a prechilled mortar and pestle with regular additions of liquid N_2 to keep the samples frozen. After the fly powder was degassed, it was homogenized 1:2 with freshly prepared DXB (25 mM Hepes pH 6.8, 50 mM KCl, 1 mM $MgCl_2$, 10% glycerol, 1 mM DTT, 1× complete protease inhibitor tablet (Roche), 0.1 U/µl RiboLock RNase Inhibitor (Fermentas), 50 mM sodium fluoride, 2 mM sodium orthovanadate, 2mM sodium pyrophosphate) with 10 strokes using the loose pestle and 10 strokes with the tight pestle in a 50 ml dounce homogenizer (Wheaton). The embryo extract was aliquoted, frozen in liquid nitrogen and stored at -80°C.

2.2.3.4.2. Immunoprecipitation and Westernblot

The frozen *Drosophila* embryo extracts were thawed on ice and centrifuged for 30 min at 4500 g. The SN was collected and centrifuged a second time. The SN of the second spin was filtered through a 5 mm diameter Schleicher & Schüll paper filter to remove the fat debris cushion, floating on top of the homogenate. The filtered extract was pre-cleared with 1ml equilibrated sepharose beads slurry for 1.5 h.

The extract was split into 2 aliquots. 1 aliquot was incubated for 1.5 h with 200 ul protein A-sepharose 4 Fast Flow beads slurry (Amersham Biosciences), preincubated O/N with 10 µg of affinity purified rabbit anti-Miranda antibody, raised against the N-terminal peptide sequence: 96C SLPQRLRFRPTPSHTDTATGSGS 118AA (Davids biotechnology). As a control, the other aliquot was incubated with beads that were pre-incubated with 10 µg of Rb IgG (Calbiochem). The beads were washed 6 times for 15 min per wash step. If the IP was performed to isolate co-precipitated RNAs, the beads were split after the last wash step.

One aliquot of the beads was boiled in SDS sample buffer and loaded on two 12% polyacrylamide gels. 1/1500 volume of the *Drosophila* embryo extract was loaded as input. For the Miranda blot, the antibody was diluted 1:300 and for the Staufen blot, the antibody was diluted 1:3000. The second aliquot of the beads was further processed for the RT-PCR reaction (see 2.2.2.3.).

For the identification of novel interacting proteins, the total amount of beads was boiled in SDS loading dye and analysed by SDS-PAGE followed by SYPRO Ruby protein staining. Protein bands present in the Miranda IP fraction but absent in the rabbit IgG control fraction, were excised and identified by mass spectrometry analysis (Prof Chris Turck, Max-Planck Institute for Psychiatry, Munich).

2.2.3.5. Sucrose Gradient

Preparation of *Drosophila* Embryo Extract

Drosophila embryo extract was prepared as described in 2.2.3.4.1 with the exception that it was prepared with DXB without glycerol and it was not frozen in liquid nitrogen after homogenization. One half of the extract was prepared with buffer containing 40U/μl Ribolock RNAse Inhibitor, whereas the other half of the extract was treated with 25 μg/ ml RNAse A. The homogenates were centrifuged for 20 min. at 10000 g in a SS34 rotor. The protein concentration of each SN was determined and 2-3 mg protein was loaded per gradient.

Preparation of 10 %-50% Sucrose Gradients

A 10 % and 50 % sucrose solution was prepared by dissolving the sucrose in DXB, prepared without glycerol. 5 ml of the 50% sucrose solution was poured in vials suitable for the SW40-Ti rotor. Carefully 5 ml of the 10% sucrose solution were poured on top. The vials were sealed and carefully positioned horizontally for 2 h at 4°C, so that the linear gradient can be established. The gradients were then carefully repositioned vertically and 2-3 mg of the *Drosophila* embryo extract was loaded per gradient. The gradients were centrifuged for 18 h at 38000 rpm, 4°C in a SW 40 rotor. After the centrifugation, 1 ml fractions of the gradients were collected from the bottom of the vial. The fractions were TCA precipitated and the protein pellets were boiled in 1× SDS loading dye. The samples were loaded on an 8 % SDS protein gel and a westernblot was performed after the proteins separated. Antibodies were diluted as described in 2.2.3.4.2.

2.2.3.6. Gelfiltration

The gelfiltration column contained Superose 6 substrate and had a bed volume of 25 ml. The substrate was equilibrated O/N with DXB (- glycerol). About 3-4 g *Drosophila* embryos were freshly homogenized in 10 ml DXB (- glycerol) and centrifuged for 20 min at 4500 g. The SN was filtered through a Schüll Paper filter. Subsequently, the SN was further filtered through a 0.45 µm and then through a 0.22 µm sterile filter. 225 µl of the filtered homogenate were injected into a 500 µl loop. The maximum pressure in the column was adjusted to 1.5 mPa. The flow rate was 100 µl/ min at the beginning and was then adjusted to 400 µl/min. 500 µl fractions were collected and further processed as described in 2.2.3.5.1.

2.2.4. Immunostaining and *In situ* Hybridization

2.2.4.1. *Drosophila* Embryo Staining

Embryos were collected and bleached as described in chapter 2.2.1.2, followed by a fixation in a 1:1 solution of 37% formaldehyde and heptane for 4 min. with vigorous shaking. The lower phase was removed as far as possible and 1 volume of methanol was added. The embryos were vortexed for 30 sec. to remove the vitelline membrane. The upper and lower phases were removed and the embryos were washed twice with methanol. After that, the embryos were shortly rinsed in PBT (1× PBS, 0.1% Triton-X 100), followed by 5 washes, 5 min. each in PBT. The embryos were incubated for at least 30 min. in blocking solution (PBT containing 5% normal goat serum (NGS)).
The primary antibodies were diluted in blocking solution and, depending on the antibodies, incubated for at least 1 h- O/N with the embryos. Subsequently they were washed 6 times for 10 min. each in PBT, followed by incubation in the secondary antibody solution (secondary fluorescently labeled antibodies diluted 1:200 in blocking solution) for 1 h. After 6 supplementary washes for 10 min. each, the embryos were incubated in TOTO®-3 iodide (Molecular probes), diluted 1:2000in 1× PBS. They were then mounted in VECTASHIELD® mounting medium on object slides.

2.2.4.2. *In situ* Hybridization

After bleaching, the embryos were fixed in 400 µl Fixation Solution (0.1 M Hepes pH 6.9, 2 mM $MgSO_4$, 1 mM EGTA, H_2O_{DEPC}), 50 µl of 37% formaldehyde, and 800 µl heptane for 20 min. The lower phase was removed as far as possible and 500 µl methanol were added. The embryos were vigorously vortexed for 30 sec to remove the vitelline membrane. The upper and lower phases were removed and the embryos were washed twice with methanol. After that, the embryos were shortly rinsed in PBT (1× PBS, 0.1% Triton-X 100, H_2O_{DEPC}), followed by 3 washes for 5 min each in PBT. The embryos were re-fixed for 15 min. in PBT with 4% formaldehyde. Finally, the embryos were washed 5× 5 min. in PBT.

Proteinase K was diluted in PBT to a final concentration of 3 µg/ ml (0.09 U/ ml) and 500 µl were added to the embryos. They were incubated for 2 min. at RT and then transferred on ice for an additional hour (the proteinase K digestion step was omitted when sensitive primary antibodies for protein co-staining were used).

The proteinase K digestion was stopped by adding PBT containing 2 mg/ ml glycine, which was removed after 2 min. This step was repeated once more and the embryos were rinsed in PBT to remove residual glycine.

The embryos were fixed again for 20 min. in PBT containing 4% formaldehyde. They were washed 5× 5 min in PBT to remove all traces of fixative, followed by a 10 min. wash step in Hybridization Solution (5× SSC pH 5.0, 50% formamide, 0.1% Tween 20, 50 µg/ ml heparin, 50 µg/ ml sonicated salmon sperm DNA, in H_2O_{DEPC}) diluted 1:1 in PBT. A final wash step for 10 min in undiluted hybridization solution followed.

100 µl of hybridization solution including 200 ng DIG labeled RNA probe were heated at 80°C for 3 min, cooled on ice for 5 min and then added to the embryos. Hybridizations were carried out at 56°C O/N in a water bath. The embryos were washed twice in hybridization solution at the hybridization temperature, followed by washes in serial dilutions (4:1, 3:2, 2:3, 1:4) of hybridization solution in PBT for 10 min. each at room temperature.

The embryos were incubated in blocking solution (PBT, 5% NGS) for 30 min., followed by an incubation in the primary antibody solution O/N (sheepαDIG antibody diluted 1:500 or mouseαDIG antibody diluted 1:250 (both from Roche)). After 6 wash steps for 10 min. each in PBT, the embryos were incubated in a solution containing the appropriate fluorescently coupled secondary antibody. Unbound antibodies were removed in 6 subsequent washing steps (10 min. each), followed by an incubation in TOTO®-3 iodide (Molecular probes), diluted 1:2000 in 1× PBS, to visualize the DNA. The embryos were then mounted in VECTASHIELD® mounting medium on object slides.

3. Results

3.1. Identification of Novel Miranda Protein Interaction Partners

3.1.1. Expression and Purification of GST-Miranda

In order to investigate protein interaction partners, involved in basal anchoring and localization of the Miranda protein, GST pull-down assays were carried out.

The N-terminal part of Miranda (amino acid 1 to 298) is sufficient to form basal crescents in mitotic *Drosophila* neuroblasts (Broadus et al, 1998). It was therefore used as bait in the GST pull-down experiments.

The cDNA encoding for residue 1 to 298 of Miranda, was cloned into the PGEX-4T1 vector. The construct was then transformed into BL21 (DE3) to perform the protein expression. After they reached an optic density of 0.6-0.8, the cells were induced by addition of IPTG. The expression was carried out for 4h at 37°C. The cells were harvested by centrifugation and broken by sonification. To control the protein expression and purification, fractions corresponding to each step were collected and analysed by SDS-PAGE (Figure 9).

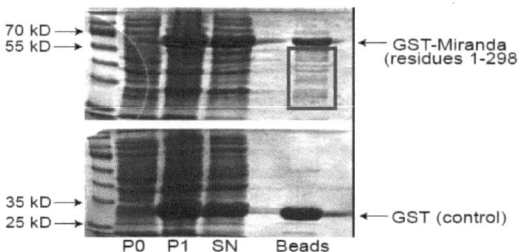

Figure 9 GST and GST-Miranda (1-298) expression and purification, analyzed by SDS-PAGE
P0 represents the protein fraction before induction by IPTG, whereas P1 represents the protein fraction 4 h after induction with IPTG. SN represents the soluble protein fraction after lysis and centrifugation. Red rectangle in the upper gel marks protein degradation products/ contaminating proteins. Gel was stained by Coomassie.

The appearance of a 55 kDa protein band in fraction P1 (after IPTG induction), which is absent in the P0 fraction (before IPTG induction) shows that the expression of the GST-Miranda fusion protein resulted from the induction with IPTG (Figure 9, upper panel, P1 compared to P0). The same result was obtained for the GST alone, which serves as a control in the following GST pull-down experiments (Figure 9, bottom panel, P1 compared to P0). The predominant bands of 55 kD and 27 kD in the GST-Miranda and GST supernatant (SN) fraction, respectively, show that the proteins were soluble.

The Miranda fusion proteins were pre-coupled to glutathione sepharose beads before they were subjected to GST pull-down experiments. The prepared beads were analyzed by SDS-PAGE (Figure 9, beads fraction). Both proteins (GST-Miranda and GST alone) bound significantly to the glutathione sepharose beads, although some contaminants were co-purified with the GST-Miranda construct (red rectangle, Figure 9, upper panel).

3.1.2. GST-Pull-Down Assays

In order to search for novel proteins that might be involved in the basal localization and cortical association of the Miranda protein complex, GST-pull-down experiments using the N-terminal protein domain as bait, were carried out. As control experiment, GST alone was used.

The GST-Miranda (1-298) fusion protein (pre-coupled to sepharose beads), was incubated with whole *Drosophila* embryo extract. After 4 h of incubation, the GST beads were washed and the bound proteins as well as the GST bait proteins were eluted by high salt treatment, TCA precipitated and analysed by SDS-PAGE.

In fact, several polypeptides were co-purified specifically with Miranda (Figure 10).

The bands were extracted from the gel and analyzed by mass spectrometry (Prof. Chris Turck, MPI for Psychiatry in Munich).

46   RESULTS

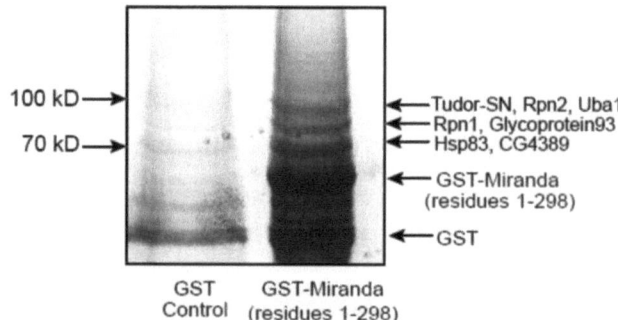

Figure 10. GST-Miranda (residues 1-298) pull-down experiment.
The eluted and TCA precipitated protein fractions from a GST pull-down experiment were analysed by SDS-PAGE. The bands extracted from the gel and analysed by mass spectrometry are marked by arrows. The proteins are visualized by SYPRO Ruby.

From the experiment, shown in Figure 10, the following proteins that specifically co-purified with Miranda were identified: Tudor-SN, CG4389, Glycoprotein 93, Heat shock protein 83 (Hsp83), Uba1, Rpn1 and Rpn2.
Several experiments of this type were performed and all the identified proteins are listed in table 2.

Name	Annotation symbol	Function	Reference
Tudor-SN	CG7008	Binding and cleavage of hyper-edited dsRNA	(Scadden, 2005)
Headcase	CG15532	Branching inhibitor in the trachea	(Weaver & White, 1995)
	CG4389	Involved in fatty acid beta oxidation	(Freeman et al, 2003)
Glycoprotein 93	CG5520	Protein folding	(Maynard, 2008)
Rpn1	CG7762	Proteolysis	(Kurucz et al, 2002)
Rpn2	CG11888	Proteolysis	(Kurucz et al, 2002)
Ubiquitin activating enzyme 1 (Uba1)	CG1782	Activates and transfers ubiquitin to ubiquitin conjugating enzymes	(Lee et al, 2008)

Heat shock protein 83 (Hsp83)	CG1242	Protein folding, intracellular signalling pathways	(Young et al, 2001)
eIF3-S10	CG9805	Translation initiation	(Andersen & Leevers, 2007)
Tripetidyl-peptidase II	CG3991	Proteolysis	(Seyit et al, 2006)
Isoleucyl-tRNA synthetase	CG11471	Isoleucyl-tRNA aminoacylation	(Seshaiah & Andrew, 1999)
Elongation factor 2b	CG2238	Translation elongation factor activity	(Lasko, 2000)
Lamin	CG6944	Nuclear membrane organization	(Goldberg et al, 1998)

Table 2. List of identified proteins that co-purified with GST-Miranda (1-298) in GST pull-down experiments.

Most of the proteins that were identified during the pull-down experiments are involved in protein degradation, in translational processes or in the nuclear membrane organization. These proteins are frequently found to bind unspecifically in protein purifications from cell extracts. Therefore they were excluded as Miranda interaction candidates.

Little is known about the protein CG4389, except that it is involved in fatty acid beta oxidation. Therefore no further analyses were pursued.

In contrary, Tudor-SN and Headcase seemed to be promising identified candidates from the GST pull-down experiments and therefore were further analysed.

Tudor-SN (Tudor Staphylococcus nuclease) corresponds to a subunit of the RNA-induced silencing complex (RISC), where it was shown to promote the cleavage of hyper edited double stranded RNA (Scadden, 2005). The name of Tudor-SN refers to the presence of a Tudor domain and five staphylococcal/ micrococcal nuclease domains in the protein.

Headcase is an extremely basic (pI 9.6) cytoplasmic protein with no obvious sequence similarities or conserved motifs in other organisms. Headcase was shown to act in an inhibitory signalling mechanism to determine the number of cells that will form unicellular sprouts in the *Drosophila* trachea (Steneberg & Samakovlis, 2001).

Headcase is also expressed in clusters of cells in the CNS during embryogenesis (Steneberg & Samakovlis, 2001).

Although there are no obvious reasons involving of Tudor-SN and Headcase in the basal anchoring or localization of Miranda or in neuroblasts, it was nevertheless interesting to further examine their potential interaction with Miranda.

3.1.3. GST Pull-Down Candidate Analyses

3.1.3.1. Tudor-SN

In order to confirm the interaction between Miranda and Tudor-SN biochemical and immunohistochemical experiments were performed.

For a biochemical approach, co-immunoprecipitation experiments, using an antibody that recognizes specifically the N-terminus of Miranda, were performed.

Practically, the antibody was incubated with whole *Drosophila* embryo extracts and Miranda containing complexes were isolated with Protein-A-Sepharose beads. The bound proteins were denaturated by boiling the beads in SDS loading dye, separated by SDS-PAGE and analysed by westernblot.

Interestingly, the westernblot in Figure 11 A shows a co-precipitation of Tudor-SN with Miranda, which confirms the result from the GST pull-down experiment.

Furthermore, FMR1 (*Drosophila* Fragile X Protein), which has been shown to exist in the same RISC complex as Tudor-SN (Caudy et al, 2003), was co-purified as well (Figure 11 A). Nevertheless, these co-precipitations were not very reproducible. Furthermore, immunoprecipitations of Tudor-SN showed only weak Miranda signals (Figure 11 B).

These results suggest transient Miranda –Tudor-SN interactions.

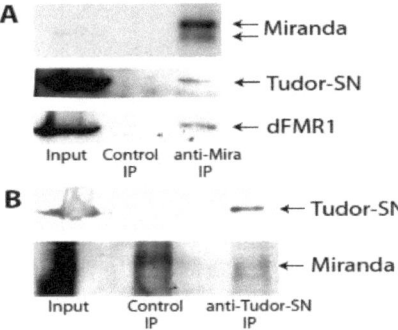

Figure 11. Miranda and Tudor-SN immunoprecipitation experiments, analysed by westernblot.
Miranda (A) and Tudor-SN (B) immunoprecipitation experiments. Tudor-SN, as well as dFMR could be co-precipitated with an anti-Miranda antibody but not with control IP (A, middle and bottom panel).

Another approach to examine an interaction between Tudor-SN and Miranda was to perform a co-staining of the two proteins in wild type *Drosophila* neuroblasts and look if they co-localize (Figure 12).

Figure 12 shows a metaphase neuroblast. Miranda forms a basal crescent, whereas Tudor-SN shows a uniform cytoplasmic distribution, which persists throughout the cell cycle (data not shown).

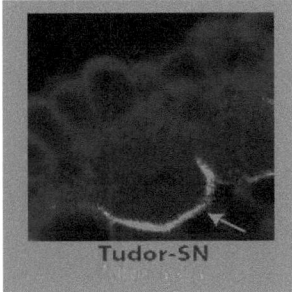

Figure 12. Tudor-SN and Miranda immunostaining in wild type embryonic neuroblasts.
Tudor-SN (blue) and Miranda (green) were stained in *Drosophila* neuroblasts. The confocal image shows a neuroblast at metaphase, where Miranda localizes to a basal crescent. White arrow indicates the neuroblast.

The Tudor-SN/ Miranda co-stainings (Figure 12) could not confirm a co-localization of the two proteins in *Drosophila* neuroblasts.

Although the obtained biochemical and immunohistochemical data could not confirm the existence of stable Tudor-SN containing Miranda complexes in *Drosophila* neuroblasts, the existence of transient forms of these complexes in neuroblasts cannot be excluded.

Nevertheless, the immunostainings excluded a possible role of Tudor-SN in localizing or anchoring Miranda to the cortex. Therefore we focused on examining the second identified candidate from the GST pull-down experiments, namely Headcase.

3.1.3.2. Headcase

In a first approach to confirm the interaction of Headcase and Miranda, I performed immunoprecipitation experiments by isolating Miranda containing complexes from *Drosophila* embryo extracts. The co-precipitated proteins were eluted and analyzed for the presence of Headcase by westernblot. Indeed, Headcase could be specifically co-precipitated with Miranda (Figure 13)

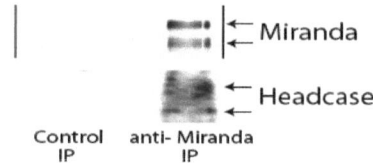

Figure 13. Miranda immunoprecipitation, analyzed by westernblot.
Headcase could be specifically co-precipitated with the Miranda antibody, but not with the IgG control, from *Drosophila* embryo extracts.

Nevertheless, like for Tudor-SN, this interaction could not continuously be reproduced.

Headcase was only be shown to be expressed in embryos beginning from stage 13 (neuroblasts start delaminating from stage 8-11 and are mitotically active to stage 14-16) (Weaver & White, 1995). This excluded Headcase in the forefront of being part of the common Miranda complex, which is expressed in all neuroblasts. Nevertheless, if Miranda and Headcase form transient complexes, this might explain the discontinuous detection of Headcase in Miranda immunoprecipitations.

In a parallel approach, Miranda immunostainings in *headcase* mutant embryos were performed (Figure 14).

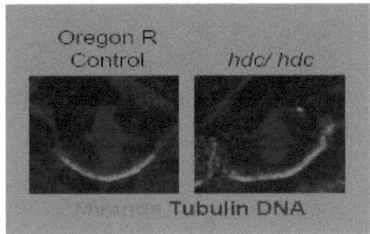

Figure 14. Miranda immunostainings in wild type (Oregon R) and *headcase* mutant neuroblasts.
Miranda localizes to a basal crescent in *headcase* mutant (*hdc/hdc*) and wild type embryonic neuroblasts at metaphase. Miranda is shown in green, tubulin in red and the DNA in blue.

Several *headcase* mutant embryos of different stages were examined, but no Miranda localization defect in *Drosophila* neuroblasts could be observed. Miranda formed normal basal crescents at metaphase (Figure 14, right panel).

Although the obtained biochemical results indicate an interaction between Miranda and Headcase, the *headcase* mutant analysis could exclude its involvement in Miranda localization or cortical association in neuroblasts.

3.1.4. Immunoprecipitation Experiments

In a further approach to identify proteins that are associated with the Miranda complex, immunoprecipitation experiments were performed.

To do that, *Drosophila* whole embryo extract was first incubated with an antibody directed against the N-terminus of Miranda and IgG of the same species, as control. Subsequently the extract was incubated with Protein-A Sepharose beads for 3 h. After the beads were washed, bound proteins were denaturated by boiling the beads directly in SDS loading dye. The proteins were analysed by SDS-PAGE (Figure 15).

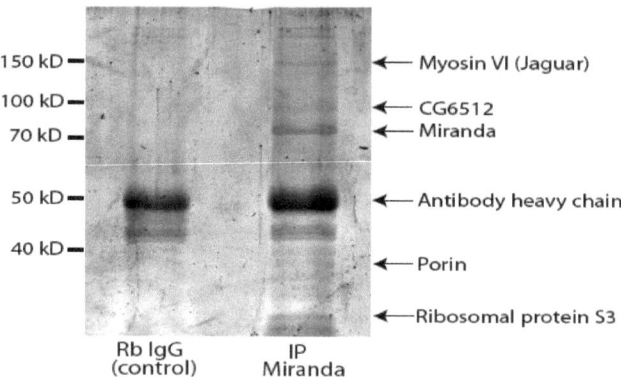

Figure 15. Miranda immunoprecipitation.
Fractions of a rabbit anti-Miranda and rabbit (Rb)-IgG control immunoprecipitation were analyzed by SDS-PAGE and stained by SYPRO Ruby. The bands indicated by the arrows were excised and identified by mass spectrometry.

The bands corresponding to the proteins that specifically co-purified with Miranda were extracted from the gel and were analysed by mass spectrometry. The following proteins could be identified in this experiment: Myosin VI, CG6512, Miranda, Porin and Ribosomal protein S3.

Several immunoprecipitation experiments were performed and the proteins, which could be identified by mass spectrometry, are listed in table 2.

Name	Annotation symbol	Function	Reference
Myosin II (Zipper)	CG15792	Non-muscle myosin	(Barros et al, 2003)
Myosin VI (Jaguar)	CG5695	Pointed end-directed myosin	(Petritsch et al, 2003)
Pavarotti	CG1258	Mitotic kinesin like protein	(Adams et al, 1998)
Paramyosin	CG5939	Major structural protein of thick filaments in invertebrate muscles	(Liu et al, 2003)
α-actinin	CG4376	Constituent of actin cytoskeleton	(Dubreuil & Wang, 2000)
Porin	CG6647	Mitochondrial porin	(De Pinto et al, 1989)
Ribosomal protein S3	CG6779	Structural constituent of ribosome	(Wilson et al, 1994)
Heat shock protein 83 (Hsp83)	CG1242	Protein folding, intracellular signalling pathways	(Young et al, 2001)
Heat shock protein cognate 3 and 4 (Hsc3, Hsc4)	CG4147 CG4264	Coordinate sequential binding and release of misfolded proteins	(Dorner et al, 2006)
eIF3-S8	CG4954	Translation initiation factor activity	(Andersen & Leevers, 2007)
Elongation factor 2b	CG2238	Translation elongation factor activity	(Lasko, 2000)
Lamin	CG6944	Nuclear membrane organization	(Goldberg et al, 1998)
	CG30015	Unknown function	

Table 2. List of identified proteins from Miranda immunoprecipitation experiments.

It is quite striking that Myosin VI (Figure 16) and Myosin II could be identified in the Miranda immunoprecipitation experiments, as these interactions have been reported previously (Petritsch et al, 2003).

This validates the strategy and the established immunoprecipitation conditions. Furthermore it allowed a certain degree of confidence for the obtained results.

As it is expected for experiments performed from whole cell extracts, several unspecifically binding proteins were identified. Like in the GST pull-down experiments, several proteins involved in translational and protein folding processes were identified. Since they represent typical unspecific binding proteins, they were excluded from further analysis. Similarly, the proteins CG6512, Porin, α-actinin and Paramyosin have no obvious link to Miranda. Therefore no further studies on these proteins were performed.

Interestingly we could identify Pavarotti. It is a kinesin-like protein, related to mammalian MKLP-1 (mitotic kinesin like protein-1). *Pavarotti* mutants exhibit defects in the embryonic nervous system (Adams et al, 1998).

As already two motor proteins were shown to be involved in Miranda's asymmetric localization (Myosin II and Myosin VI), it was really promising to identify this candidate.

3.1.5. Pavarotti Analyses

To examine a possible requirement of Pavarotti to localize or anchor Miranda in *Drosophila* neuroblasts, immunostainings in *pavarotti* mutant embryos were performed (Figure 16).

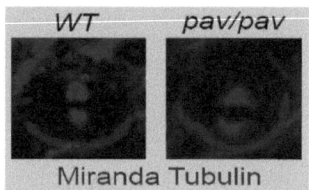

Figure 16. Miranda immunostainings in wild type and *pavarotti* mutant neuroblasts.
Miranda localizes normally in *pavarotti* mutant (*pav/pav*) neuroblasts. Confocal images of metaphase neuroblasts with Miranda in red and Tubulin in blue.

Detailed examination of Miranda in several *pavarotti* mutant embryos could not reveal a localization defect in neuroblasts (Figure 16).
Biochemical approaches to confirm the interaction between Pavarotti and Miranda were not successful (data not shown). Therefore, the existence of Miranda complexes containing Pavarotti could not be confirmed.

In summary I could identify Tudor-SN, Headcase and Pavarotti and Myosin VI in the protein interaction experiments. Unfortunately, further analyses revealed that the Tudor-SN and Headcase interactions with Miranda seem to be transient.

The obtained results raised the possibility that Miranda might exist in different complexes in *Drosophila*. To elucidate this aspect, experiments to characterize Miranda complexes biochemically, were performed.

3.2. Biochemical Characterization of Miranda Complexes

3.2.1. Linear 10%- 50% Sucrose Gradient

An approximate size determination of Miranda complexes was performed with a linear 10%- 50% sucrose gradients.
Therefore, freshly prepared *Drosophila* whole embryo extract was split into two aliquots. One aliquot was treated with RNAse and the second with RNAse inhibitor. The intention of this RNAse treatment is, to allow a size distinction between Miranda protein and Miranda protein/ RNA containing complexes. This is interesting, as Miranda was shown to transport *prospero* mRNA (via Staufen) in neuroblasts.
Drosophila embryo extract (3 mg protein, adjusted to 500 µl volume with extraction buffer, see materials and methods) of either fraction was then loaded onto a linear 10% to 50% sucrose gradient. A gel filtration standard (Thyroglobulin 670 kD, γ-Globulin 158 kD, Ovalbumin 44 kD) was loaded on a separate gradient to control the migration. The sucrose gradients were subjected to centrifugation for 18 h, 4°C, at 30000 g. Fractions were collected, TCA precipitated and analysed by westernblot (Figure17).

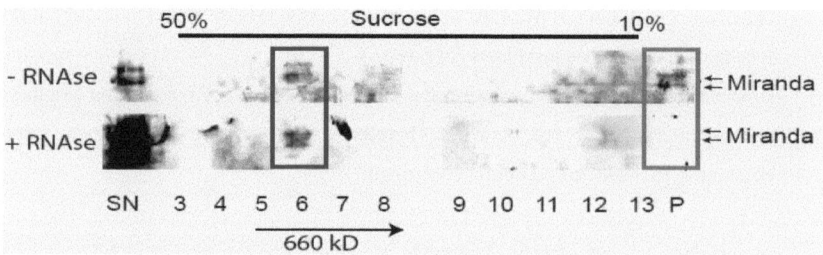

Figure 17. Westernblot analysis of 10%-50% sucrose gradient fractions.
Upper panel shows fractions of the sucrose gradient with non-RNAse treated extract analysed by westernblot, whereas the bottom panel shows fractions of the sucrose gradient with RNAse treated extract. Red rectangle marks Miranda complexes of approximately 660 kDa, whereas the green rectangle marks Miranda complexes found in the pellet (P).

Interestingly, the sucrose gradients reveal the existence of Miranda complexes of approximately 660 kDa in presence or absence of RNAse (Figure 17, red rectangle). Therefore these might correspond to RNAse insensitive Miranda protein complexes.

The estimation of 660 kDa could correspond to the theoretical molecular weight (MW) of Miranda and its identified cargo proteins (Miranda ~ 90 kDa, Staufen ~ 110 kDa, Brat ~ 110 kDa, Prospero ~ 200 kDa and Myosin VI (Jaguar) ~ 140 kDa).

Moreover, Miranda was shown to be co-localized with *prospero* mRNA (via Staufen) throughout the cell cycle in neuroblasts. The *prospero* mRNA has a calculated molecular weight of 2.3 MDa. Therefore it does not seem surprising to detect a Miranda signal in the RNAse sensitive pellet (Figure 17, green rectangle).

These experiments indicate for the first time the existence of at least two Miranda complex populations: RNAse insensitive complexes (corresponding to 660 kDa) and RNA containing complexes (higher than 2 MDa).

To exclude the possibility that the 660 kDa complexes, result from instability of Miranda complexes in high concentrations of sucrose, gelfiltration experiments were performed.

3.2.2. Gelfiltration

Westernblot analysis of gelfiltration fractions could indeed confirm the presence of two Miranda complex populations (Figure 18).

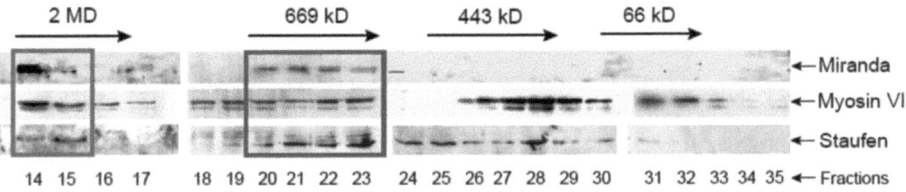

Figure 18. Westernblot analysis of Superose-6 gelfiltration fractions.
Miranda, Myosin VI and Staufen Westernblot analysis of TCA precipitated gelfiltration fractions. Arrows above the westernblots indicate the migration of protein complexes of a gelfilration standard. The green rectangle marks complexes with a molecular weight of at least 2 MDa, whereas the red rectangle marks complexes of approximately 660 kDa.

The red rectangle marks complexes of approximately 669 kDa, whereas the fractions marked by the green rectangle correspond to complexes of 2 MDa or higher (Figure 18). This indicates that the smaller Miranda containing complexes (669 kDa) do not result from dissociation due to the sucrose gradient conditions. Apparently, the identified Miranda interaction partners Staufen and Myosin VI exist with Miranda in both complexes. The used gelfiltration matrix was Superose-6, and the column had a size exclusion of 2 MDa. Therefore, this experiment did not allow an approximate size estimation of high molecular weight Miranda complexes presumably containing RNA, as *prospero* was calculated to 2.3 MDa.

Thus, these data provide evidence for the existence of at least two Miranda containing complexes. One complex of an approximate molecular weight of 660 kDa is insensitive to RNAse treatment. In contrary, the other complex that has an approximate molecular weight of at least 2 MDa, shows sensitivity to RNAse.

3.3. Identification of Novel RNAs, Associated to Miranda Complexes

Headcase and Tudor-SN could be identified as binding partners of Miranda in immunoprecipitation experiments. It could also be shown that Miranda is part of at least two complexes. One of them is RNA sensitive. This result encouraged us to further investigate Miranda´s role in transporting RNAs.

3.3.1. Miranda Immunoprecipitation and Candidate PCR Analysis

In order to identify novel mRNAs, associated with Miranda, the complex was immunoprecipitated under non-denaturing and RNAse-free conditions from *Drosophila* whole embryo extracts.

Different groups of RNA candidates were tested for their ability to be co-purified with Miranda (table 3). One group was composed of genes encoding for proteins which are localized in the *Drosophila* neuroblast, like e.g. *staufen* or *par-6*. Another group contained genes, from which the corresponding RNAs were shown to be expressed in neuroblasts/ GMCs in a published screening for novel neural precursor genes (Brody et al, 2002). These candidates included e.g. *dacapo*.

In addition, other candidates like the genes involved in the microRNA pathway (*dicer-1* and *argonaute-1*) were tested, because it was shown before that germline stem cell division in *Drosophila* is controlled by the microRNA pathway (Hatfield et al, 2005).

The specific immunoprecipitation conditions, where established for the genes *inscuteable* and *prospero*. *Inscuteable* RNA persists apically in the neuroblast throughout the cell cycle and the proteins required for its localization do not include Staufen or Miranda (Hughes et al, 2004). Therefore, *inscuteable* was considered as negative control. *Prospero*, which was shown to be transported by Miranda served as positive control.

Miranda could be precipitated using a specific N-terminal antibody (Figure 20 A, Miranda IP). In the control experiment, no immunoprecipitation of Miranda using Rb IgG was observed (Figure 20 A, control IP).

WB analysis revealed a co-precipitation of Staufen with Miranda (Figure 20A). RT-PCR analysis of IP fractions followed by candidate PCR shows that *dacapo* RNA is associated with Miranda (Figure 20 B, top panel). *Prospero* and *inscuteable* served as positive and negative control, respectively.

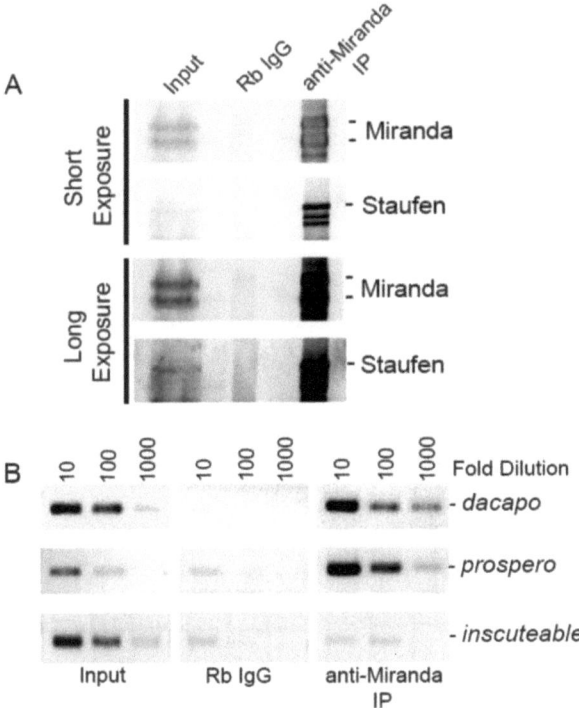

Figure 20. Miranda immunoprecipitation and candidate PCR analysis.
Anti-Miranda antibody specifically precipitates Miranda and its known cargo protein Staufen from *Drosophila* embryo extracts (A). After the co-immunoprecipitated RNA was submitted to reverse transcription, *dacapo* as well as *prospero* (positive control) were specifically detected in the Miranda but not IgG co-precipitate by PCR (B, top and middle panel). *Inscuteable* RNA, which is known to be apically localized in the neuroblast throughout the cell cycle, was precipitated with neither Miranda antibody nor IgG and serves as a negative control (B, bottom row).

All positive and negative candidates that were obtained from the Miranda immunoprecipitation experiment, followed by candidate PCR analysis, are shown in Figure 21.

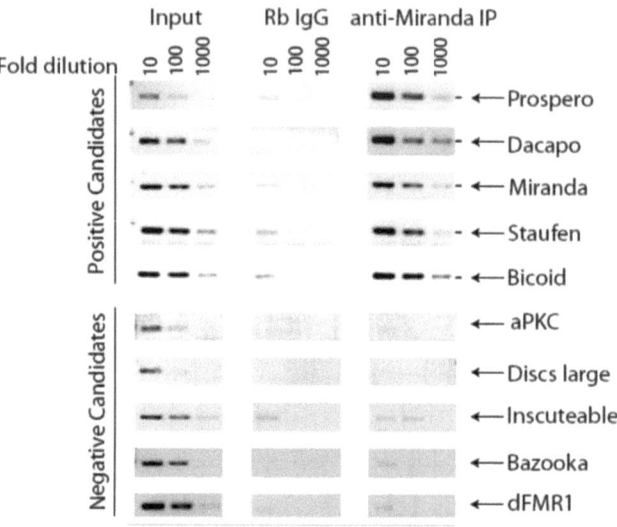

Figure 21. Candidate PCR analysis of Miranda immunoprecipitation fractions.
Template DNA for the PCR analysis was obtained from reverse transcription of RNA, which co-precipitated with Miranda. The DNA was diluted 1:10, 1:100 and 1:1000 for the PCR reaction.

Prospero, *dacapo*, *staufen*, *miranda* and *bicoid* were positive candidates, meaning that they show clearly elevated signal intensity in the fraction obtained from the Miranda IP in contrast to the signal observed from in the Rb IgG control fraction (Figure 21).
The candidates, which showed comparable signal intensities in the anti-Miranda IP as well as in the Rb IgG control fraction were considered as negative (data not shown).

The candidates that were tested on the RNA pools of several performed immunoprecipitation experiments are listed in table 3.

Name	Annotation Symbol	Function
Prospero	CG17228	RNA localized by Miranda, positive control
Miranda	CG12249	Protein asymmetrically localized in neuroblast
Inscuteable	CG11312	RNA persists apically, negative control
Lgl	CG2671	Protein asymmetrically localized in neuroblast
Bazooka	CG5055	Protein asymmetrically localized in neuroblast
G alpha I	CG10060	Protein asymmetrically localized in neuroblast
Par-6	CG5884	Protein asymmetrically localized in neuroblast
Pon	CG3346	Protein asymmetrically localized in neuroblast
Numb	CG3779	Protein asymmetrically localized in neuroblast
Staufen	CG5753	Protein asymmetrically localized in neuroblast
Pins	CG5692	Protein asymmetrically localized in neuroblast
Dlg	CG1725	Protein asymmetrically localized in neuroblast
aPKC	CG10261	Protein asymmetrically localized in neuroblast
Crumbs	CG6383	Protein expressed in neuroblast
Hunchback	CG9786	Transcription factor, expressed in neuroblast
Castor	CG2102	Transcription factor, expressed in neuroblast
Krueppel	CG3340	Transcription factor, expressed in neuroblast
Mushroom-body expressed	CG7437	(Brody et al, 2002)
Hnr27C	CG10377	(Brody et al, 2002)
Split ends	CG18497	(Brody et al, 2002)
Failed axon connections	CG4609	(Brody et al, 2002)

Blistery	CG9379	(Brody et al, 2002)
Mab-2	CG4746	(Brody et al, 2002)
Dappled	CG1624	(Brody et al, 2002)
Hs2st	CG10234	(Brody et al, 2002)
ImpL3	CG10160	(Brody et al, 2002)
	CG5358	(Brody et al, 2002)
Nerfin 1	CG13906	(Brody et al, 2002)
	CG7372	(Brody et al, 2002)
Myb	CG9045	(Brody et al, 2002)
Dref	CG5853	(Brody et al, 2002)
Mcm7	CG4978	(Brody et al, 2002)
CDC45L	CG3658	(Brody et al, 2002)
Set	CG4299	(Brody et al, 2002)
Adar	CG12598	(Brody et al, 2002)
Elav	CG4262	(Brody et al, 2002)
Drumstick	CG10016	(Brody et al, 2002)
Notch	CG3936	Expressed in neuroblast
CDC2	CG5363	Cell cycle
Cyclin E	CG3938	Cell cycle
Neuralized	CG11988	(Brody et al, 2002)
	CG13920	(Brody et al, 2002)
Schizo	CG10577	(Brody et al, 2002)

	CG5235	(Brody et al, 2002)
	CG5358	(Brody et al, 2002)
dFMR	CG6203	MicroRNA pathway
Dicer 1	CG4792	MicroRNA pathway
Argonaute 1	CG6671	MicroRNA pathway
Dacapo	CG1772	(Brody et al, 2002)

Table 3. List of candidate genes, tested for Miranda association.

From all tested candidates, only *dacapo* and *prospero* could be repeatedly co-precipitated.

Dacapo encodes the *Drosophila* CIP/KIP-type cyclin dependent kinase inhibitor, specific for Cyclin E/ Cdk2 complexes (de Nooij et al, 1996; Lane et al, 1996).

This was a very promising finding, as the neuroblast and the GMC have different requirements in term of cell cycle factors. A neuroblast can divide many times without differentiating and therefore resembles a stem cell, whereas the GMC only divides once to generate neurons or glia. It seems quite plausible that Miranda might contribute to these intrinsic differences not only by transporting cell fate determinants (Prospero and Brat) to the GMC, but also by transporting the RNA of cell cycle regulators such as *dacapo*.

The positive presence of *bicoid* in few experiments, is consistent with a publication, showing that Miranda has the ability to interfere with the Staufen/ *bicoid* localization pathway in early embryos (Irion et al, 2006).

Miranda and *staufen* were not regularly co-precipitated in the performed experiments.

In contrary the candidate *dacapo*, which repeatedly co-precipitated with Miranda, was clearly considered for further examination.

3.3.2. Dacapo in situ Hybridization Experiments

In order to confirm the result, obtained from the candidate PCR analysis, I performed whole mount *in situ* hybridization experiments to detect *dacapo* RNA in *Drosophila* embryos. A co-staining of *dacapo* RNA and Miranda protein would clarify if they are co-localized in neuroblasts.

To perform the *in situ* hybridizations, digoxigenin labeled RNA probes were generated. *Prospero* and *inscuteable* probes were generated for control experiments.

The quality and the quantity of the probes after *in vitro* transcription were evaluated by native agarose gel electrophoresis (Figure 22).

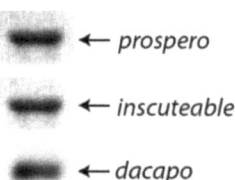

Figure 22. Digoxigenin labeled RNA probes.
Digoxigenin labeled RNA probes of *prospero*, *inscuteable* and *dacapo* were analysed by native agarose gel electrophoresis after in vitro transcription.

The first step towards examining a possible co-localization of *dacapo* and Miranda was to establish the conditions for the *in situ* hybridizations, combined with Miranda protein staining. This was performed by recapitulating the published *prospero* and *inscuteable* expression pattern. It has been shown that *prospero* localizes to a basal crescent in metaphase, whereas *inscuteable* persists apically throughout the cell cycle (Hughes et al, 2004; Li et al, 1997).

Figure 23 confirms the reported RNA expression data of *prospero* and *inscuteable* (Hughes et al, 2004; Li et al, 1997) and shows for the first time a co-staining of Miranda protein. *Prospero* co-localizes with Miranda to a basal crescent (Figure 23, upper panels), whereas *inscuteable* persists mainly apically in metaphase neuroblasts (Figure 23, bottom panels).

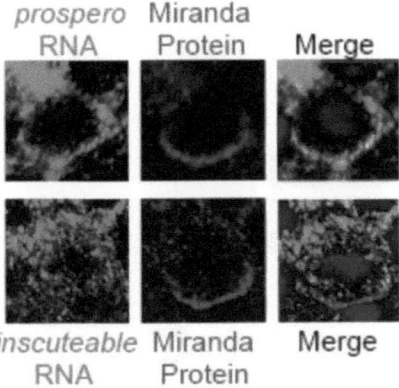

Figure 23. *In situ* hybridization of *prospero* and *inscuteable* with Miranda protein staining. Confocal images of metaphase neuroblasts, stained for *prospero* RNA (in green) and Miranda protein (in red, upper panel), as well as for *inscuteable* RNA (in green) and Miranda (red, bottom panel). DNA is stained in blue.

Since the appropriate conditions for a co-staining of RNA and protein were established, the examination of *dacapo* could be performed.

Interestingly, *dacapo* showed a co-localization with Miranda, Prospero and Staufen throughout the cell cycle. Figure 24 A shows *dacapo* co-localized with Miranda and Prospero to a basal crescent in metaphase. Figure 24 B demonstrates the co-localization of *dacapo* with Staufen at different phases of the cell cycle. In interphase/ prophase, Staufen and *dacapo* are co-localized to an apical crescent. They both form a basal crescent in metaphase and are inherited by the GMC in telophase.

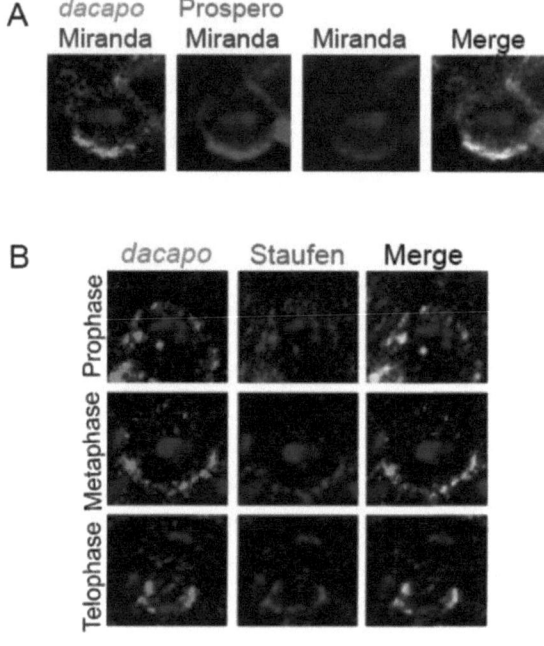

Figure 24. *In situ* **hybridization of *dacapo* with Miranda/ Prospero and Staufen protein staining.**
Confocal images of a metaphase neuroblast, showing *dacapo* (green) co-localized with Miranda (blue) and Prospero (red) to a basal crescent (A). Pro-, meta- and anaphase neuroblast showing *dacapo* (green) colocalized with Staufen (red) throughout the cell cycle (B). DNA is stained in blue in A and B.

In order to test the hypothesis that *dacapo* localization in *Drosophila* neuroblasts could be Staufen dependent, as it was shown for *prospero* (Broadus & Doe, 1997), experiments on *staufen* mutant embryos were performed (Figure 25).

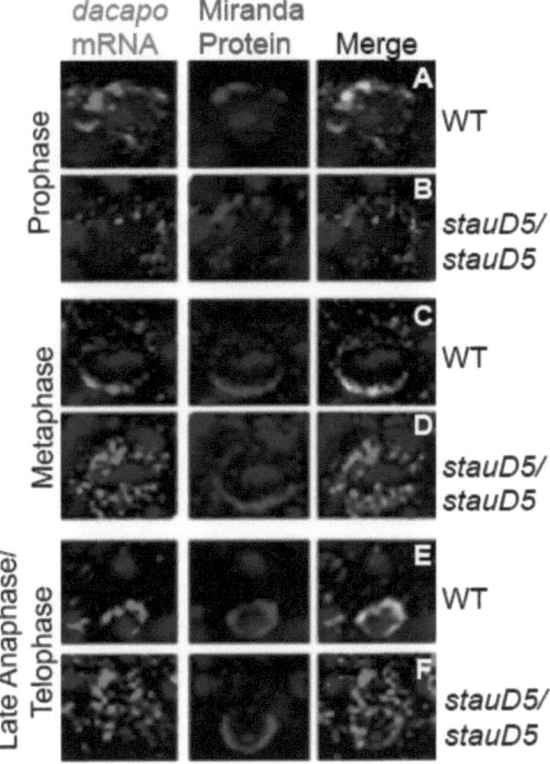

Figure 25. *In situ* hybridizations of *dacapo* and Miranda protein staining in wild type and *staufen* mutant embryos.
Confocal images of wild type (WT) and *staufen* mutant (*stauD5/ stauD5*) embryos. They were stained for *dacapo* RNA (green) and Miranda protein (red). Representative neuroblasts in pro-, meta- and telophase are shown in A/ B, C/ D and E/ F respectively. DNA is stained in blue.

In fact, whole mount *in situ* hybridizations of *dacapo* with Miranda protein staining could show that *dacapo* RNA is mislocalized in the absence of Staufen, whereas Miranda localizes normally (Figure 25).

Miranda is localized apically in interphase/ prophase (Figure 25 A and B), forms a basal crescent in metaphase (Figure 25 C and D) and is inherited by the GMC in anaphase/ telophase (Figure 25 E and F). Whereas *dacapo* co-localizes with Miranda throughout the cell cycle in WT embryos (Figure 25 A, C and E), it is mislocalized to the cytoplasm in *staufen* mutant neuroblasts throughout the cell cycle (Figure 25 B, D and F).
Staufen contains five copies of double stranded RNA (dsRBD) binding motifs (St Johnston et al, 1992).

Staufen protein plays an important role in anterior-posterior axis formation during *Drosophila* oogenesis, by localizing *oskar* mRNA to the posterior pole of the oocyte where the abdomen and the germline will form (Ephrussi et al, 1991; Kim-Ha et al, 1991; St Johnston et al, 1991) and by localizing *bicoid* mRNA to the anterior pole after the egg has been laid (Ferrandon et al, 1994; St Johnston et al, 1989).
The dsRBD2 of all Staufen homologues is split by a proline-rich insertion in one of the RNA-binding loops. A deletion of this insertion reveals a role for this dsRBD in the localization of *oskar* mRNA, whereas *prospero* localization is not affected {Micklem, 2000 #35}. Removal of dsRBD5 (*stauD5/stauD5*) leads to a disruption of proper *prospero* localization in the neuroblast, but the *oskar* mRNA localizes normally even though it is not translated at the posterior of the oocyte.
Indeed it has been shown that Staufen binds directly to Miranda via the dsRBD5 and thereby couples Staufen/ *prospero* mRNA complexes to the actin-based localization pathway in neuroblasts (Broadus et al, 1998; Schuldt et al, 1998).
Therefore, a fly strain expressing the Staufen protein that lacks the dsRBD5 in a *staufen* mutant background (a kind gift of D. St. Johnston, Wellcome Trust, Cambridge UK) was used for the *dacapo* mislocalization analysis in Figure 25.

This experiment could clearly show that *dacapo* mRNA localization by Miranda in *Drosophila* is Staufen dependent and that it requires the same dsRBD as *prospero*.
I could observe that *dacapo* was not expressed in all neuroblasts. Specifically, *dacapo* was mainly observed in neuroblast of older embryos, meaning that it was not expressed at stages when the neuroblasts start to delaminate. This was the next step to elucidate.

3.3.3. Quantification of Miranda/ *Dacapo* Co-Expressing Neuroblast Sizes

In order to verify the observation that *dacapo* and Miranda are mainly co-expressed in neuroblasts of later embryonic stages, quantifications were carried out.

These quantifications compared the plane sizes of Miranda/ *dacapo* co-expressing neuroblasts to stage specific neuroblast planes (Figure 26).

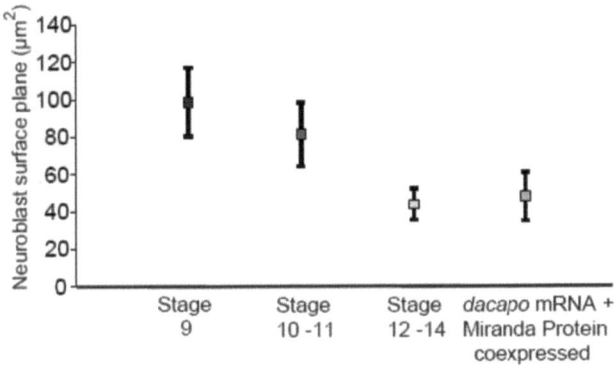

Figure 26. Quantification of Miranda and *dacapo* coexpressing neuroblasts.
A quantification of neuroblast planes at different embryonic stages (according to Hartenstein), revealed a co-expression of Miranda and *dacapo* RNA at late embryonic stages. At stage 9 neuroblasts showed a mean surface plane of 98.51±18.25 µm^2 (n=93), at stage 10-11 the mean plane was 81.18±17.09 (n=95) and at stage 12-14 it was 43.54±8.51 (n=26). Neuroblasts co-expressing *dacapo* RNA and Miranda protein showed a mean surface plane of 47.54±13.11 (n=49) consistent with stage 10/11 and later neuroblasts.

The surface planes were measured with the quantification tool, provided with the Leica SP2 confocal microscope software.

At stage 9, neuroblasts showed a mean surface plane of 98.51±18.25 µm^2 (n=93) at stage 10-11 the mean plane was 81.18±17.09 (n=95) and at stage 12-14 it was 43.54±8.51 (n=26). Neuroblasts co-expressing *dacapo* RNA and Miranda protein showed a mean surface plane of 47.54±13.11 (n=49).

70 RESULTS

The obtained data correspond to neuroblasts of embryonic stages 10/ 11 and later. These findings are very consistent with the published *dacapo* expression pattern in the central nervous system (Le Borgne et al, 2002).

3.3.4. Dacapo RNA and Protein Staining

I wanted to clarify, if the Dacapo protein expression correlates to the localization of the RNA. It should be expected, that the asymmetric RNA localization would result in the appearance of the protein in the GMC, but not in the neuroblast.

Co-stainings of Dacapo protein and RNA revealed a different pattern. In fact, *dacapo* RNA is localized apically in prophase, forms a basal crescent in metaphase and is inherited exclusively by the GMC in telophase (Figure 27 A), whereas the Dacapo protein is diffusely localized to the cytoplasm in the neuroblast and is equally distributed to both daughter cells upon telophase (Figure 27 A).

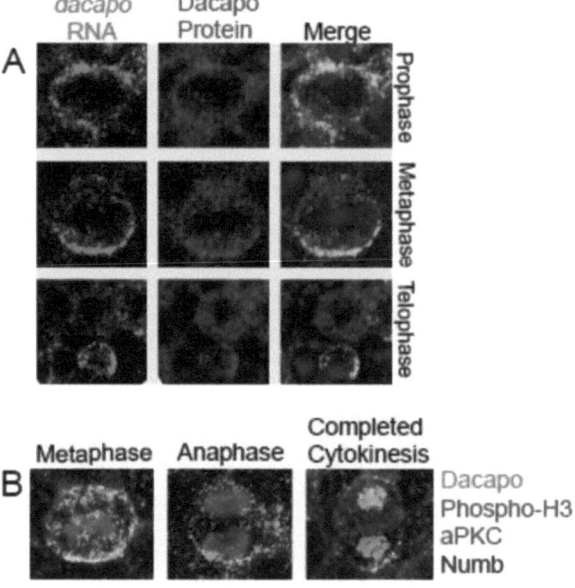

Figure 27. Dacapo protein distribution in *Drosophila* neuroblasts.
Confocal images of *Dacapo* RNA (green) and Dacapo protein (red), co-stained in pro-, meta- and telophase neuroblasts (A). B shows Dacapo protein (green) co-stained with the asymmetrically localized proteins aPKC (red) and Numb (blue) (B). DNA is stained in blue (A) or in red (B).

The specificity of the Dacapo expression pattern in neuroblasts was verified by stainings of Dacapo protein with aPKC and Numb, two asymmetrically localized proteins in the neuroblast. Whereas aPKC persists at the apical cortex throughout the cell cycle, and is inherited by the neuroblast daughter, Numb is localized from the apical cortex in prophase to the basal cortex in metaphase and is inherited by the GMC daughter cell.

Indeed, Dacapo protein is mainly localized to the cytoplasm in *Drosophila* neuroblasts and is inherited by both daughter cells which can be seen in the last image of Figure 28 B. There Dacapo protein locates to the nuclei of both daughter cells, to function as inhibitor of Cyclin E/ Cdk2 complexes.

These results display a different Dacapo protein expression than it is expected from the RNA localization. Whereas *dacapo* RNA is asymmetrically localized in the neuroblast and exclusively inherited by the GMC daughter cell, the protein is present in the neuroblast and it is inherited by both daughter cells.

3.3.5. Dacapo Mutant Analyses

The most obvious question that evolves at this stage is what role Dacapo plays in neuroblasts / GMCs. Therefore, *dacapo* mutant analyses were performed. To do that, I performed BrdU (bromodeoxyuridine) labeling experiments of wild type and *dacapo* mutant embryos.

3.3.5.1. BrdU Labeling

BrdU is a synthetic nucleoside, which is an analogue of thymidine. It is commonly used for the detection of proliferating cells in living tissues.

Living embryos were incubated in a BrdU containing solution for 30 min. and then subjected to immunostaining (Figure 28).

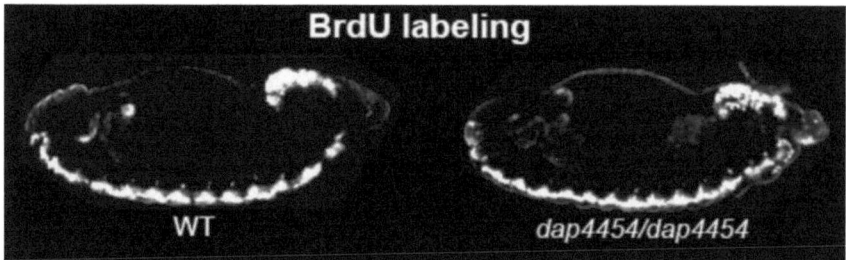

Figure 28. BrdU labeling of *Drosophila* embryos.
Confocal images of stage 13 *Drosophila* embryos. Wild type (wt) and *dacapo* mutant (*dap4454/dap4454*) were stained for incorporation of BrdU.

Dacapo acts as cyclin dependent kinase inhibitor specific for CyclinE/ Cdk2 complexes. This results in cells arresting at G1. It was therefore examined, if *dacapo* mutant embryos show additional mitotic activity in the CNS.
Although it was not possible to detect minor cell changes from images taken at this resolution, *dacapo* mutant embryos did not show clearly elevated mitotic activity in the CNS compared to wild type embryos (Figure 28).
To exclude the possibility that additional cells *dacapo* mutants are eliminated by higher apoptotic activity and therefore are not apparent in the BrdU labeling, immunostainings for cleaved Caspase-3 were performed.

3.3.5.2. Caspase-3 Staining

Caspase-3 is one of the key players of apoptosis. The activation of caspase-3 requires its cleavage into activated fragments. The used antibody recognizes one of these activated fragments and is therefore an excellent detection tool for apoptotic cells.

Wild type and *dacapo* mutant embryos of different embryonic stages where stained for apoptotic activity. Figure 29 shows stage 12/13 embryos that were stained for cleaved Caspase-3.

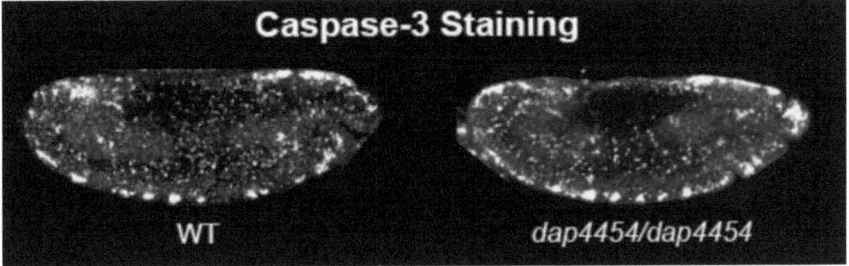

Figure 29. Immunostaining of apoptotic cells in wild type and *dacapo* mutant embryos.
Confocal images of stage 12/13 *Drosophila* embryos. Wild type and *dacapo* mutant (*dap4454/dap4454*) embryos were stained for cleaved Caspase-3 which indicates apoptotic cells.

No obvious differences in the appearance of apoptotic activity could be observed by comparing wild type to *dacapo* mutant embryos.

Nevertheless, the comparison was limited by the fact that significant variations among individual embryos in terms of apoptotic activity existed.

Dacapo is required for final mitosis of the embryonic epidermis. As epidermal cells divide synchronously and all cells arrest at the same time, *dacapo* mutant analysis by BrdU labeling could clearly visualize an additional division of all epidermal cells (de Nooij et al, 1996; Lane et al, 1996).

Drosophila neuroblasts not only delaminate asynchronously, they also show extreme lineage specific differences in the number of resulting progeny cells.

Therefore it is probably more promising to examine *dacapo* mutant phenotypes specifically in individual neuroblast lineages.

3.3.5.3. Dacapo Mutant Analysis in the Neuroblast 6-4 Lineage

A mutation in *dacapo* was shown to result in slight numerical abnormalities in the progeny of the specific neuroblast lineage 6-4 (NB 6-4) (Berger et al, 2005).
The NB 6-4 lineage can be easily recognized by expression of the marker protein Eagle (Eg) that is also expressed in the neuroblast lineages 2-4, 3-3, 7-3 and by a specific number and localization pattern of neurons and glia cells. The number and assembly pattern varies from the thoracic (T) segments to the abdominal (A) segments (Figure 30 A). Whereas 3 glia cells are arranged on either side of the midline in the thoracic segments, only 2 glia cells are formed at either side of the midline in the abdominal segments (Figure 30 A and B, left panels). *Dacapo* mutant embryos show one additional glia cell in the abdominal segments of NB 6-4 (Figure 30 A and B, left panels).

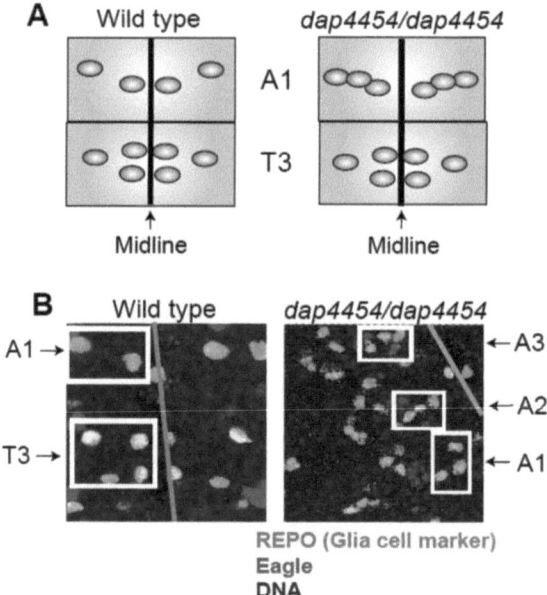

Figure 30. *Dacapo* mutant analysis in the neuroblast lineage 6-4.
Schematic representations of glia cell arrangement in the thoracic (T3) and abdominal (A1) segments of the neuroblast lineage 6-4 (A). Confocal images of stage 14 embryos (B). Wild type and *dacapo* mutant (*dap4454/dap4454*) embryos are stained for the glia cell progeny of the neuroblast 6-4 lineage. The glia cell marker REPO is stained in green. Eagle, marking the neuroblast lineages 6-4, 2-4, 3-3 and 7-3 is stained in red and DNA is stained in blue.

The white rectangles mark the glia cell clusters of NB6-4, represented in the schemes above. The orange bar marks the midline.

The obtained data from the *dacapo* mutant analysis in the NB-6-4 lineage (Figure 30) are consistent with published results (Berger et al, 2005).

In summary it could be demonstrated that Miranda is required for the asymmetric localization of *dacapo* RNA in *Drosophila* neuroblasts. The RNA is co-localized with Miranda throughout the cell cycle and is exclusively segregated to the GMC. Furthermore, it could be shown that this asymmetric localization is Staufen dependent. The RNA starts to be expressed in neuroblasts of later staged embryos and Dacapo influences the number of cell divisions in at least one neuroblast lineage. These results confirm the role of Miranda not only for asymmetric localization of proteins, but also for localizing RNAs.

4. Discussion

The general goals of this thesis were to perform an initial biochemical characterization of Miranda containing complexes and to identify novel proteins and RNAs that associate with Miranda.

Biochemical characterizations of Miranda containing complexes by sucrose gradients and gelfiltrations could reveal the existence of at least 2 Miranda complex populations. One population is insensitive to RNAse treatment and has an approximate size of 660 kDa. At least one other population was detected that showed sensitivity to RNAse treatment and has a molecular weight of at least 2 MDa.

The protein interaction partner search was carried out with GST pull down experiments and immunoprecipitations, followed by an identification of the co-purified proteins by mass spectrometry. The obtained candidates Tudor-SN and Headcase could be co-precipitated with Miranda in immunoprecipitations. Nevertheless they do not seem to be part of the common identified Miranda complexes in *Drosophila* neuroblasts.

The trials to find novel RNAs that associate with Miranda resulted in the identification of *dacapo*. This RNA shows a co-localization with Miranda in neuroblasts throughout the cell cycle. Interestingly, this RNA localization depends on the dsRNA binding protein Staufen, which is also responsible for the localization of *prospero,* an identified Miranda associated RNA. *Dacapo* encodes the *Drosophila* CIP/KIP-type cyclin dependent kinase inhibitor, specific for Cyclin E/ Cdk2 complexes (de Nooij et al, 1996; Lane et al, 1996). The obtained results suggest a role of Dacapo in arresting cell proliferation in neuroblast lineages.

4.1. Identification of Novel Miranda Protein Binding Partners

The aim of the search for novel binding partners was to identify proteins that play a role in Miranda's localization and cortical association in neuroblasts.

Miranda acts as an adaptor protein for Prospero, Staufen and Brat and co-localizes with its cargoes throughout the cell cycle in *Drosophila* neuroblasts. Furthermore, *prospero* mRNA is constantly co-localized with Miranda by direct binding to Staufen. Miranda binds to the apical cortex in prophase, localizes to the basal cortex in metaphase and is then inherited by the GMC in telophase.

The motor proteins Myosin II and Myosin VI were shown to be involved in the localization of Miranda (Barros et al, 2003; Petritsch et al, 2003). Myosin II acts by excluding Miranda from binding to the apical cortex after prophase, whereas Myosin VI co-localizes with Miranda in the neuroblast cytoplasm.

Although they were both shown to bind directly to Miranda, Myosin II and Myosin VI do not exist in the same complex when purified from embryonic extracts. Therefore the presence of a shuttle protein that performs the transport Miranda from one complex to another seemed possible. Furthermore, the basal cortical anchor protein for Miranda was not discovered, as Myosin VI co localizes mainly in the cytoplasm with Miranda and therefore is excluded as candidate (Petritsch et al, 2003).

Two different strategies were pursued to identify novel interacting proteins.

One strategy was to perform GST pulldown experiments from *Drosophila* embryo extracts. The Miranda protein domain that was shown to be sufficient for its proper localization was used as bait (Fuerstenberg et al, 1998; Shen et al, 1998).

The other approach was to isolate Miranda containing complexes from embryo extracts by immunoprecipitation of Miranda. The proteins that specifically co-purified in the pulldown and immunoprecipitation experiments were analysed by mass spectrometry.

With both strategies several proteins could be co-purified. The immunoprecipitation strategy could be validated by the repeated presence of Myosin II and Myosin VI in several trials. Both motors could be identified as Miranda interacting proteins by this method (Petritsch et al, 2003).

As expected for experiments from whole cell extracts, several unspecifically binding proteins involved in transcription-/ translation- and protein degradation processes were identified. The most promising candidates from both strategies were further analysed. These were Tudor-SN, Headcase and Pavarotti.

Tudor-SN represents a subunit of RISC (RNA induced silencing complex) and it is involved in binding and promoting the cleavage of hyper edited double stranded RNA generated by ADARs (Scadden, 2005).
ADARs (adenosine deaminases that act on RNA) catalyze one type of RNA editing whereby hydrolytic delamination converts adenosine (A) to Inosine (I). Selective editing results in few A→ I conversions and is important for regulation of gene expression, although hyperediting of long perfect dsRNA by ADARs may result in up to 50% of A residues converted to I (Nishikura et al, 1991; Polson & Bass, 1994).

Westernblot analyses of Miranda immunoprecipitation fractions revealed a co-precipitation of Tudor-SN. Nevertheless, these co-precipitations could not constantly be reproduced. Miranda westernblots from Tudor-SN immunoprecipitation fractions resulted sporadically in a faint detection of Miranda. Although these results indicate the existence of Tudor-SN containing Miranda complexes, these interaction seem to be quite transient.
Immunostainings of Tudor-SN revealed a diffuse cytoplasmic distribution of the protein in *Drosophila* neuroblasts. Miranda shows no co-localization with Tudor-SN at any phase of the cell cycle. Therefore, the existence of Tudor-SN and Miranda containing complexes in *Drosophila* neuroblasts could not be confirmed.

The second putative Miranda interacting protein identified by GST pulldown, was Headcase. Headcase acts as branching inhibitor in the *Drosophila* trachea (Steneberg et al, 1998). It is involved in an inhibitory signalling mechanism that determines the number of cells that form unicellular sprouts in the trachea.
As Headcase protein was also shown to be expressed in clusters of cells in the CNS during neurogenesis (Weaver & White, 1995), it was further examined.

I performed immunoprecipitation experiments with an antibody against Miranda. Headcase could be specifically co-purified with Miranda. Nevertheless it could not constantly be detected.

Immunostainings in *headcase* mutant embryos could not reveal a localization defect of Miranda in *Drosophila* neuroblasts.
Apparently Headcase can form complexes with Miranda, but the reported expression data exclude Headcase from being part of common Miranda complexes, expressed in all neuroblasts. Although it cannot be excluded that Headcase and Miranda form complexes in specific neuroblasts, the extensive *headcase* mutant analysis could eliminate a possible role in localizing Miranda.

Miranda expression is not limited to neuroblasts. It has been reported that Miranda shows a broad pattern of expression pattern and that it also associates with centrosomes (Mollinari et al, 2002). As Miranda inherits several putative aPKC phosphorylation sites, it seems quite likely that phosphorylation modifications might influence the protein binding properties of Miranda in different cell types.
This could explain why Headcase and Tudor-SN could be co-purified with Miranda in immunoprecipitations, although the presence of these complexes could not be confirmed in neuroblasts. The irregular detection in the immunoprecipitations might result from differing abundances of the proteins in the embryonic pools that were subjected to the experiments, due to varying embryonic stages.

As two motor proteins (Myosin II and Myosin VI) have been implicated in the asymmetric localization of Miranda, it was encouraging to identify Pavarotti co-purified with Miranda in an immunoprecipitation experiment.
Pavarotti is a kinesin-like protein that is a member of the MKLP-1 (mitotic kinesin-like protein-1) subfamily. Unfortunately, further performed immunoprecipitation trials could not confirm this interaction and Miranda localized normally in *pavarotti* mutant neuroblasts. Therefore, the existence of Miranda and Pavarotti containing complexes could not be confirmed.

Pavarotti complexes are cytoplasmic at prophase, associated with mitotic spindles during metaphase, concentrated in the spindle midzone during anaphase, localized to the midbody at cytokinesis and to the nucleus during interphase (Somers & Saint, 2003). Interestingly, this localization pattern resembles the reported pattern of maternally contributed Miranda protein in early embryos.

Mollinari and collegues reported Miranda around the nuclei on the centrosomes in prophase. In metaphase and anaphase, Miranda is not only accumulated on the centrosomes at the opposite poles of the mitotic spindle, but also on the spindle itself. In telophase, Miranda re-localizes to the midbodies (Mollinari et al, 2002).

The accordance of both described expression patterns is really striking and makes the existence of Miranda and Pavarotti containing complexes in early neuroblasts likely.

In a recent publication by Erben and colleagues we could clarify aspects of apical to basal Miranda localization in neuroblasts. Apparently Miranda reaches the basal cortex by passive diffusion throughout the cell, rather than by long-range Myosin VI directed transport. Myosin VI acts by delivering diffusing Miranda to the basal cortex.

This is a major finding, because I wanted to identify proteins that are the missing link between Mirandas apical association with Myosin II and the cytoplasmic involvement of Myosin VI. It seems quite likely that Miranda localization does not require additional proteins. Nevertheless, the basal anchor of the protein complex remains unidentified.

None of Miranda's cargo proteins was identified by mass spectrometry in the experimental trials. Their presence could only be confirmed by westernblot analysis. One explanation might be that they possibly do not exist in the same stoichiometric ratios as Miranda, or only bind transiently. This would lead to a quantity that lies under the detection limit of the used protein staining solution.

A recent publication reported the first solution structure of Miranda's central cargo binding domain (CBD) (Yousef et al, 2008). The Miranda CBD forms a parallel coiled-coil homodimer (Figure 31).

It is suggested that the dimeric Miranda CBD can bind multiple cargos simultaneously on its elongated coiled-coil region and the unwound termini.

Dimerization of the two identical N-terminal domains, which can form coiled coil and the double heads of the Miranda dimer (Figure 31), might be required to increase the affinity for specific binding partners.

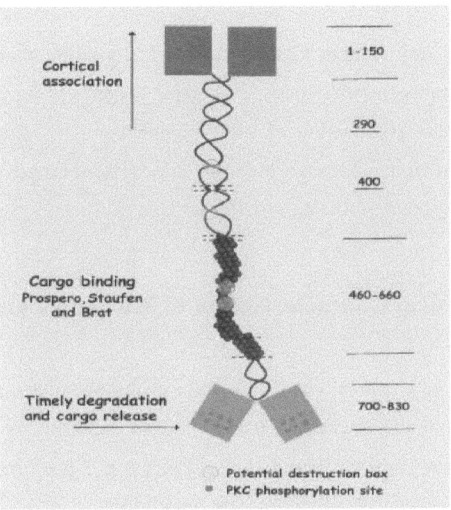

Figure 31 Structure model of the Miranda protein.
Representation of full-length Miranda. Residue numbers are shown on the right. Green circles at the C-terminal region represent phosphorylation sites. Gray circles indicate four potential destruction boxes. Predicted non-coiled-coil regions are indicated by dotted lines. Cyan and yellow boxes represent the N and C termini (Yousef et al, 2008).

If one assumes that Miranda binds cargo as dimer, the monomeric N-termini of the GST fusion proteins were therefore probably not able to bind interacting proteins with sufficient high affinity.

Therefore future GST pull-down experiments should include the CBD in the Miranda fusion baits, so that dimerization is possible and the identification of novel interaction partners is more likely.

Further approaches to identify novel interacting proteins, especially the basal anchor protein, might include yeast two hybrid assays. These assays already led to the identification of Prospero (Shen et al, 1997), Staufen (Schuldt et al, 1998) and most recently to Brat (Lee et al, 2006), as binding partners of Miranda. The verification of positive candidates could then be performed by co-IPs and immunostainings in neuroblasts.

Furthermore a very evident tool for further Miranda interaction studies would be the generation of transgenic fly strains, expressing tagged versions of Miranda, which would allow the affinity purification of these complexes. This strategy also resulted in the identification of Brat as Miranda binding partner (Betschinger et al, 2006).

4.2. Biochemical Characterization of Miranda Complexes

Until now, no biochemical characterization of the Miranda complex has been reported. The results from the interaction partner search experiments indicate the existence of several Miranda complex populations. Therefore I performed sucrose gradient and gelfiltration experiments to evaluate this possibility.

The experiments revealed for the first time the presence of RNAse sensitive and RNAse insensitive Miranda complexes.

RNAse sensitive complexes had an estimated molecular weight of at least 2 MDa. They probably correspond to the reported Miranda complexes in *Drosophila* neuroblasts, because *prospero* RNA was shown to be co-localized with Miranda throughout the cell cycle and we could calculate the molecular size of this RNA to 2.3 MDa.

In contrast, the presence of the estimated 660 kDa RNAse insensitive complexes is surprising. The RNAse insensitive complex of 660 kDa existed not only in the extract treated with RNAse, but also which was not treated with RNAse and even protected from possible degradation by addition of RNAse inhibitor.

Unfortunately the size of the RNA containing Miranda complexes could not be estimated with the performed experimental approaches.

Although Gelfiltration experiments are a commonly used tool for a rough size estimation of protein or protein/ nucleic acid complexes, the reliability of this method is limited, because it only works under the presumption that all macromolecules behave uniform, which likely is not the case for complexes containing nucleic acids.

Highest accuracy in size determination is obtained by subjecting the complexes to native mass spectrometry, dynamic light scattering or/ and analytical ultracentrifugation.

A prerequisite for the named methods however, is the quality and quantity of the samples to be examined.

The initial biochemical approaches on Miranda complexes paved the way to perform further experiments in this direction. It could clearly be shown that Miranda exists in different complexes and it would be interesting to determine the proteins of each complex and correspond the complexes to their cellular localization.

4.3. Identification of Novel, to Miranda Complexes Associated RNAs

As complementary approach to characterize Miranda complexes, it was examined, if further RNAs are associated.

Up to now, *prospero* RNA is the only identified RNA that was described to be transported by Miranda via the RNA-binding adaptor Staufen (Broadus et al, 1998; Schuldt et al, 1998)

An advantage of localizing the mRNA instead of the protein, is the fact that the transcript can facilitate many rounds of protein synthesis., which avoids the energy cost of moving each protein molecule individually (Jansen, 2001).

As Miranda has a major role in transporting cell fate determinating proteins into the GMC, it seemed promising to explore its role as adaptor for transporting RNAs.

Therefore Miranda complexes were isolated from *Drosophila* embryo extracts under RNAse free conditions by immunoprecipitation. The isolated RNAs were reverse transcribed and the resulting cDNA was subjected to candidate PCR analysis.

The candidates for the PCR analysis were selected due to the role their corresponding proteins have in the neuroblast or GMC (Brody et al, 2002).

I could repeatedly detect *dacapo* specifically co-associated with the immunoprecipitated Miranda complex.

In order to examine an *in vivo* co-localization of *dacapo* and Miranda in the *Drosophila* neuroblast, I performed whole mount *in situ* hybridization (ISH) experiments.

The *dacapo* ISH experiments in combination with Miranda protein staining revealed a co-localization throughout the cell cycle. Miranda as well as *dacapo* was inherited exclusively by the GMC after cell division.

Furthermore, I could demonstrate that the *dacapo* localization is Staufen dependent. *Dacapo* was mislocalized to the cytoplasm in embryos that express a mutated from of *staufen*, lacking the dsRBD5. This binding domain of Staufen was previously shown to be required for proper localization of *prospero* RNA (Broadus et al, 1998).

Dacapo was identified in 1996, as *Drosophila* CIP/KIP-type cyclin dependent kinase inhibitor, specific for Cyclin E/ Cdk2 complexes (de Nooij et al, 1996; Lane et al, 1996). Dacapo has a highly dynamic expression pattern in *Drosophila* embryos and in each of its appearance, *dacapo* RNA expression seems to coincide with cell cycle arrest during development (de Nooij et al, 1996).

Dacapo is required for final mitosis of the embryonic epidermis. In *dacapo* mutant embryos, epidermal cells undergo one extra cell cycle. The G1 arrest which is observed after the terminal division of the epidermal cells is dependent on the inactivation of Cyclin E/ CDK2 activity. In addition to the upregulation of Dacapo, the down-regulation of Cyclin E seems to contribute to the timely inactivation of Cyclin E/ CDK2 activity (Knoblich et al, 1994). This downregulation of Cyclin E activity could also be observed in *dacapo* mutant epidermal cells.

Neuroblasts divisions give rise to a further neuroblast and a GMC daughter cell. As GMCs divide only once more to generate neurons or glia cells, it seemed an attractive idea that *dacapo* might be expressed in all GMCs to prevent further proliferation after its division.

On the basis of quantifying the plane sizes of neuroblasts, I could show that *dacapo* only starts to be expressed in neuroblasts of stage 10/11 embryos and is absent in early GMCs. These results show that cell cycle exit in the early CNS is not dependent on Dacapo.

The role of Dacapo in the CNS/ neuroblast has not been elucidated yet. Therefore I performed initial *dacapo* mutant analysis.

Several distinct neuroblast lineages exist, which delaminate at specific embryonic stages and exhibit specific division features. The results from the epidermal cells (one extra cell cycle) lead to the presumption that *dacapo* phenotypes show rather slight numerical variations.

In tissues with asynchronously dividing cell lineages, these examinations therefore have to be more specified to a single cell level. In the case of neuroblasts this means the examination of a specific neuroblast lineage.

It was reported that Dacapo influences the number of progeny cells in the specific neuroblast lineage NB6-4 in *Drosophila* embryos (Berger et al, 2005).

The thoracic neuroblast lineage 6-4 (NB6-4t) generates both neurons and glia cells, whereas the abdominal neuroblast lineage 6-4 (NB6-4a) generates only glia cells. The NB6-4t lineage represents the ground state, whereas the NB6-4a lineage is specified by the homeotic genes Abdominal A (Abd-A) and Abdominal B (Abd-B). This specification takes place by down-regulating levels of CycE, which is asymmetrically expressed after the first division of NB6-4t.

Therefore I examined the NB6-4 lineage in the thoracic and abdominal segments of *dacapo* mutant embryos and compared it to wild type. Indeed, *dacapo* mutant embryos show an additional glia cell in NB6-4 in the abdominal segments, whereas the glia cell number in the thoracic segments is unchanged. These data are consistent with the previously published report from Berger and colleagues (Berger et al, 2005).

Although it could be revealed that Dacapo influences the number of progeny cells in a neuroblast lineage, it exhibits segment specific phenotypes.

The process of neuroblast delamination has been divided into five successive waves (S1-S5) with particular subpopulations of identified NBs delaminating during each wave. Each neuroblast expresses a specific set of molecular markers (Doe, 1992).

The size of the neuroblast clones produced during the embryonic phase of neurogenesis varies immensely: at one extreme the neuroblast MP2 generates only two cells (Bossing et al, 1996), whereas neuroblast NB7-1 can produce more than 40 cells (Schmid et al, 1999).

Interestingly Dacapo expression could be observed in the MP2 neuroblast. This is a special type of neuroblast that does not express Miranda and shows Prospero accumulated in the nucleus (Meyer et al, 2002).

The neuroblast divides only once to produce two postmitotic neurons (Doe et al, 1988; Spana & Doe, 1995). Therefore the MP2 neuroblast resembles more a GMC than a neuroblast. Furthermore, MP2 is comparable with a neuroblast that ceases division, as it divides only once.

I presume that Dacapo occupies the same role in the neuroblasts that it has in the epidermis, namely that it starts to be expressed when the neuroblast are about entering a G1 state. This coincides with the downregulation of Cyclin E and other cell cycle regulators.

This idea is emphasized by the above mentioned finding that Dacapo exists in the cytoplasm of the GMC like/ ceasing like neuroblast MP2.

The reason why *dacapo* RNA is asymmetrically localized into the GMC is not clear yet, especially as Dacapo protein can be found in the cytoplasm of the same neuroblasts. Probably the RNA serves as a back-up mechanism for the protein, as it was assumed for *prospero* RNA.

Prospero protein like its RNA, is asymmetrically localized by Miranda and only enters the nucleus in the GMC (except MP2 neuroblast), although it is also expressed in the neuroblast. It has been shown to bind upstream of over 700 genes, many of which are involved in neuroblast self-renewal or cell-cycle control. Furthermore it can also induce the expression of neural differentiation genes (Choksi et al, 2006). The possible role of Prospero as transcriptional activator and/or inhibitor might assure proper cell cycle exit after the GMC divided.

To obtain further insights of Dacapo´s function in the CNS, future experiments should include the generation MARCM (mosaic analysis with a repressible cell marker) clones. That would allow a more specific visualization of Dacapo´s influence on neuroblast divisions. Furthermore it needs to be characterized, which sequences in the *dacapo* RNA are needed for its asymmetric localization.

The performed experiments could not completely exclude the possibility that the other tested candidates in the PCR analysis are absent from Miranda complexes, especially the candidates we did not further analyze due their erratic specific appearances in the PCR analysis.

For future identifications of Miranda associated RNAs, it would clearly be recommended to perform Microarray experiments of Miranda IP vs. Miranda control fractions. With the performed immunoprecipitation strategy it was not possible to obtain RNA of the quality required for Microarrays.

Strategies for obtaining RNA of sufficient quality might involve approaches that would allow affinity purifications of Miranda complexes (TAP, FLAG and Myc). Furthermore, trials to accumulate the target cell population (neuroblasts) in the total cell pool, from which Miranda complexes are isolated, should be performed. Possibly by preparing the cell extracts from dissected embryonic ventral nerve cords.

4.4. Conclusion and Outlook

The goals of this thesis were to identify proteins that are involved in the localization and anchoring of Miranda complexes in *Drosophila* neuroblasts and to identify further RNAs that are associated.

With the performed strategies for the protein interaction partner search, several proteins specifically co-purified and could be identified. The most promising candidates Tudor-SN, Headcase and Pavarotti were further analysed. Tudor-SN and Headcase could be co-precipitated with Miranda. Nevertheless, no relevance of these interactions in neuroblasts could be discovered. An influence of Pavarotti on Miranda localization could not be confirmed. Nevertheless, Miranda expression is not limited to neuroblasts and the reported Miranda localization pattern in early embryos strikingly resembles the pattern reported for Pavarotti. Therefore, an interaction of both might take place at early embryonic stages.

Initial biochemical approaches to characterize Miranda complexes, revealed the presence of at least 2 complex populations. One population represents RNAse insensitive complexes corresponding to ~660 kDa.

At least one other population exists, which exhibits sensitivity to RNAse treatment. Although size estimations for the high molecular weight complexes were not possible due to experimental limitations, they are likely to correspond to complexes of at least 2 MDa.

In the approach to identify further RNAs associated to the Miranda complex, I could detect *dacapo* RNA repeatedly co-purified with Miranda. *In situ* hybridizations could confirm a co-localization of *dacapo* RNA with Miranda throughout the cell cycle in neuroblasts.

Dacapo corresponds to *Drosophila* CIP/KIP-type cyclin dependent kinase inhibitor, specific for Cyclin E/ Cdk2 complexes.

Staufen mutant analysis could substantiate the hypothesis that *dacapo* localization in *Drosophila* neuroblasts relies on the same mechanism as it has been demonstrated for *prospero* before.

Neuroblast quantifications revealed that *dacapo* only starts to be expressed in stage 10/11 embryonic neuroblasts. This corroborates previous findings that *dacapo* RNA expression coincides with cell cycle arrest during development.

Mutant analysis could reveal that Dacapo influences the number of cell divisions at least in a specific neuroblast lineage.

The fact, that Miranda exists in more than one complex in *Drosophila*, could allow a multi-functional role in the embryo, beyond asymmetrically localizing cell fate determinants in neuroblasts.

Nevertheless, several aspects of Miranda function in neuroblasts remain unclear. Especially the basal anchor protein remains unidentified. Taken together, the performed experiments paved the way for several new insights into Miranda features and provide a starting point for investigating several new aspects of this versatile protein.

References

Adams RR, Tavares AA, Salzberg A, Bellen HJ, Glover DM (1998) pavarotti encodes a kinesin-like protein required to organize the central spindle and contractile ring for cytokinesis. *Genes Dev* **12**(10): 1483-1494

Adereth Y, Dammai V, Kose N, Li R, Hsu T (2005) RNA-dependent integrin alpha3 protein localization regulated by the Muscleblind-like protein MLP1. *Nat Cell Biol* **7**(12): 1240-1247

Affolter M, Barde Y (2007) Self-renewal in the fly kidney. *Dev Cell* **13**(3): 321-322

Andersen DS, Leevers SJ (2007) The essential Drosophila ATP-binding cassette domain protein, pixie, binds the 40 S ribosome in an ATP-dependent manner and is required for translation initiation. *J Biol Chem* **282**(20): 14752-14760

Ashburner M (1989) *Drosophila*: A Laboratory Handbook. *Cold Spring Harbor Laboratory Press, Cold Spring Harbor, New York*

Bardin AJ, Le Borgne R, Schweisguth F (2004) Asymmetric localization and function of cell-fate determinants: a fly's view. *Curr Opin Neurobiol* **14**(1): 6-14

Barr FA, Sillje HH, Nigg EA (2004) Polo-like kinases and the orchestration of cell division. *Nat Rev Mol Cell Biol* **5**(6): 429-440

Barros CS, Phelps CB, Brand AH (2003) Drosophila nonmuscle myosin II promotes the asymmetric segregation of cell fate determinants by cortical exclusion rather than active transport. *Dev Cell* **5**(6): 829-840

Bashirullah A, Halsell SR, Cooperstock RL, Kloc M, Karaiskakis A, Fisher WW, Fu W, Hamilton JK, Etkin LD, Lipshitz HD (1999) Joint action of two RNA degradation pathways controls the timing of maternal transcript elimination at the midblastula transition in Drosophila melanogaster. *EMBO J* **18**(9): 2610-2620

Bello B, Reichert H, Hirth F (2006) The brain tumor gene negatively regulates neural progenitor cell proliferation in the larval central brain of Drosophila. *Development* **133**(14): 2639-2648

Berger C, Pallavi SK, Prasad M, Shashidhara LS, Technau GM (2005) A critical role for cyclin E in cell fate determination in the central nervous system of Drosophila melanogaster. *Nat Cell Biol* **7**(1): 56-62

Berleth T, Burri M, Thoma G, Bopp D, Richstein S, Frigerio G, Noll M, Nusslein-Volhard C (1988) The role of localization of bicoid RNA in organizing the anterior pattern of the Drosophila embryo. *EMBO J* **7**(6): 1749-1756

Betschinger J, Knoblich JA (2004) Dare to be different: asymmetric cell division in Drosophila, C. elegans and vertebrates. *Curr Biol* **14**(16): R674-685

Betschinger J, Mechtler K, Knoblich JA (2006) Asymmetric segregation of the tumor suppressor brat regulates self-renewal in Drosophila neural stem cells. *Cell* **124**(6): 1241-1253

Bossing T, Udolph G, Doe CQ, Technau GM (1996) The embryonic central nervous system lineages of Drosophila melanogaster. I. Neuroblast lineages derived from the ventral half of the neuroectoderm. *Dev Biol* **179**(1): 41-64

Brendza RP, Serbus LR, Duffy JB, Saxton WM (2000) A function for kinesin I in the posterior transport of oskar mRNA and Staufen protein. *Science* **289**(5487): 2120-2122

Broadus J, Fuerstenberg S, Doe CQ (1998) Staufen-dependent localization of prospero mRNA contributes to neuroblast daughter-cell fate. *Nature* **391**(6669): 792-795

Broadus J, Skeath JB, Spana EP, Bossing T, Technau G, Doe CQ (1995) New neuroblast markers and the origin of the aCC/pCC neurons in the Drosophila central nervous system. *Mech Dev* **53**(3): 393-402

Brody T, Odenwald WF (2000) Programmed transformations in neuroblast gene expression during Drosophila CNS lineage development. *Dev Biol* **226**(1): 34-44

Brody T, Stivers C, Nagle J, Odenwald WF (2002) Identification of novel Drosophila neural precursor genes using a differential embryonic head cDNA screen. *Mech Dev* **113**(1): 41-59

Bullock SL, Ish-Horowicz D (2001) Conserved signals and machinery for RNA transport in Drosophila oogenesis and embryogenesis. *Nature* **414**(6864): 611-616

Campos-Ortega JA, and Hartenstein, V. (1985) *The Embryonic Development of Drosophila melanogaster.*

Caudy AA, Ketting RF, Hammond SM, Denli AM, Bathoorn AM, Tops BB, Silva JM, Myers MM, Hannon GJ, Plasterk RH (2003) A micrococcal nuclease homologue in RNAi effector complexes. *Nature* **425**(6956): 411-414

Ceron J, Gonzalez C, Tejedor FJ (2001) Patterns of cell division and expression of asymmetric cell fate determinants in postembryonic neuroblast lineages of Drosophila. *Dev Biol* **230**(2): 125-138

Choksi SP, Southall TD, Bossing T, Edoff K, de Wit E, Fischer BE, van Steensel B, Micklem G, Brand AH (2006) Prospero acts as a binary switch between self-renewal and differentiation in Drosophila neural stem cells. *Dev Cell* **11**(6): 775-789

D. SJ (1993) Pole plasm and the posterior group of genes. *Bate & Martinez-Arias*: 325-363

de Nooij JC, Letendre MA, Hariharan IK (1996) A cyclin-dependent kinase inhibitor, Dacapo, is necessary for timely exit from the cell cycle during Drosophila embryogenesis. *Cell* **87**(7): 1237-1247

De Pinto V, Benz R, Caggese C, Palmieri F (1989) Characterization of the mitochondrial porin from Drosophila melanogaster. *Biochim Biophys Acta* **987**(1): 1-7

Ding D, Parkhurst SM, Halsell SR, Lipshitz HD (1993) Dynamic Hsp83 RNA localization during Drosophila oogenesis and embryogenesis. *Mol Cell Biol* **13**(6): 3773-3781

Doe CQ (1992) Molecular markers for identified neuroblasts and ganglion mother cells in the Drosophila central nervous system. *Development* **116**(4): 855-863

Doe CQ, Goodman CS (1985) Early events in insect neurogenesis. I. Development and segmental differences in the pattern of neuronal precursor cells. *Dev Biol* **111**(1): 193-205

Doe CQ, Hiromi Y, Gehring WJ, Goodman CS (1988) Expression and function of the segmentation gene fushi tarazu during Drosophila neurogenesis. *Science* **239**(4836): 170-175

Dorner S, Lum L, Kim M, Paro R, Beachy PA, Green R (2006) A genomewide screen for components of the RNAi pathway in Drosophila cultured cells. *Proc Natl Acad Sci U S A* **103**(32): 11880-11885

Driever W, Nusslein-Volhard C (1988) A gradient of bicoid protein in Drosophila embryos. *Cell* **54**(1): 83-93

Dubreuil RR, Wang P (2000) Genetic analysis of the requirements for alpha-actinin function. *J Muscle Res Cell Motil* **21**(7): 705-713

Ephrussi A, Dickinson LK, Lehmann R (1991) Oskar organizes the germ plasm and directs localization of the posterior determinant nanos. *Cell* **66**(1): 37-50

Erben V, Waldhuber M, Langer D, Fetka I, Jansen RP, Petritsch C (2008) Asymmetric localization of the adaptor protein Miranda in neuroblasts is achieved by diffusion and sequential interaction of Myosin II and VI. *J Cell Sci* **121**(Pt 9): 1403-1414

Fehon RG, Kooh PJ, Rebay I, Regan CL, Xu T, Muskavitch MA, Artavanis-Tsakonas S (1990) Molecular interactions between the protein products of the neurogenic loci Notch and Delta, two EGF-homologous genes in Drosophila. *Cell* **61**(3): 523-534

Ferrandon D, Elphick L, Nusslein-Volhard C, St Johnston D (1994) Staufen protein associates with the 3'UTR of bicoid mRNA to form particles that move in a microtubule-dependent manner. *Cell* **79**(7): 1221-1232

Frank DJ, Edgar BA, Roth MB (2002) The Drosophila melanogaster gene brain tumor negatively regulates cell growth and ribosomal RNA synthesis. *Development* **129**(2): 399-407

Freeman MR, Delrow J, Kim J, Johnson E, Doe CQ (2003) Unwrapping glial biology: Gcm target genes regulating glial development, diversification, and function. *Neuron* **38**(4): 567-580

Fuerstenberg S, Peng CY, Alvarez-Ortiz P, Hor T, Doe CQ (1998) Identification of Miranda protein domains regulating asymmetric cortical localization, cargo binding, and cortical release. *Mol Cell Neurosci* **12**(6): 325-339

Gavis ER, Lehmann R (1992) Localization of nanos RNA controls embryonic polarity. *Cell* **71**(2): 301-313

Ghysen A, Dambly-Chaudiere C (1989) Genesis of the Drosophila peripheral nervous system. *Trends Genet* **5**(8): 251-255

Giet R, McLean D, Descamps S, Lee MJ, Raff JW, Prigent C, Glover DM (2002) Drosophila Aurora A kinase is required to localize D-TACC to centrosomes and to regulate astral microtubules. *J Cell Biol* **156**(3): 437-451

Goldberg M, Lu H, Stuurman N, Ashery-Padan R, Weiss AM, Yu J, Bhattacharyya D, Fisher PA, Gruenbaum Y, Wolfner MF (1998) Interactions among Drosophila nuclear envelope proteins lamin, otefin, and YA. *Mol Cell Biol* **18**(7): 4315-4323

Gore AV, Maegawa S, Cheong A, Gilligan PC, Weinberg ES, Sampath K (2005) The zebrafish dorsal axis is apparent at the four-cell stage. *Nature* **438**(7070): 1030-1035

Hatfield SD, Shcherbata HR, Fischer KA, Nakahara K, Carthew RW, Ruohola-Baker H (2005) Stem cell division is regulated by the microRNA pathway. *Nature* **435**(7044): 974-978

Hirata J, Nakagoshi H, Nabeshima Y, Matsuzaki F (1995) Asymmetric segregation of the homeodomain protein Prospero during Drosophila development. *Nature* **377**(6550): 627-630

Hughes JR, Bullock SL, Ish-Horowicz D (2004) Inscuteable mRNA localization is dynein-dependent and regulates apicobasal polarity and spindle length in Drosophila neuroblasts. *Curr Biol* **14**(21): 1950-1956

Hulskamp M, Schroder C, Pfeifle C, Jackle H, Tautz D (1989) Posterior segmentation of the Drosophila embryo in the absence of a maternal posterior organizer gene. *Nature* **338**(6217): 629-632

Ikeshima-Kataoka H, Skeath JB, Nabeshima Y, Doe CQ, Matsuzaki F (1997) Miranda directs Prospero to a daughter cell during Drosophila asymmetric divisions. *Nature* **390**(6660): 625-629

Irion U, Adams J, Chang CW, St Johnston D (2006) Miranda couples oskar mRNA/Staufen complexes to the bicoid mRNA localization pathway. *Dev Biol* **297**(2): 522-533

Irion U, Leptin M, Siller K, Fuerstenberg S, Cai Y, Doe CQ, Chia W, Yang X (2004) Abstrakt, a DEAD box protein, regulates Insc levels and asymmetric division of neural and mesodermal progenitors. *Curr Biol* **14**(2): 138-144

Isshiki T, Pearson B, Holbrook S, Doe CQ (2001) Drosophila neuroblasts sequentially express transcription factors which specify the temporal identity of their neuronal progeny. *Cell* **106**(4): 511-521

Ito K, Hotta Y (1992) Proliferation pattern of postembryonic neuroblasts in the brain of Drosophila melanogaster. *Dev Biol* **149**(1): 134-148

Ito K. UJ, Technau GM. (1995) Distribution, classification, and development of Drosophila glial cells in the late embryonic and early larval ventral nerve cord. *Roux's Arch Dev Biol* **204**: 284-307

Izumi Y, Ohta N, Itoh-Furuya A, Fuse N, Matsuzaki F (2004) Differential functions of G protein and Baz-aPKC signaling pathways in Drosophila neuroblast asymmetric division. *J Cell Biol* **164**(5): 729-738

Jansen RP (2001) mRNA localization: message on the move. *Nat Rev Mol Cell Biol* **2**(4): 247-256

Jarman AP, Grau Y, Jan LY, Jan YN (1993) atonal is a proneural gene that directs chordotonal organ formation in the Drosophila peripheral nervous system. *Cell* **73**(7): 1307-1321

Jimenez F, Campos-Ortega JA (1990) Defective neuroblast commitment in mutants of the achaete-scute complex and adjacent genes of D. melanogaster. *Neuron* **5**(1): 81-89

Jongens TA, Hay B, Jan LY, Jan YN (1992) The germ cell-less gene product: a posteriorly localized component necessary for germ cell development in Drosophila. *Cell* **70**(4): 569-584

Kaltschmidt JA, Brand AH (2002) Asymmetric cell division: microtubule dynamics and spindle asymmetry. *J Cell Sci* **115**(Pt 11): 2257-2264

Kambadur R, Koizumi K, Stivers C, Nagle J, Poole SJ, Odenwald WF (1998) Regulation of POU genes by castor and hunchback establishes layered compartments in the Drosophila CNS. *Genes Dev* **12**(2): 246-260

Kanai MI, Okabe M, Hiromi Y (2005) seven-up Controls switching of transcription factors that specify temporal identities of Drosophila neuroblasts. *Dev Cell* **8**(2): 203-213

Kiebler MA, Hemraj I, Verkade P, Kohrmann M, Fortes P, Marion RM, Ortin J, Dotti CG (1999) The mammalian staufen protein localizes to the somatodendritic domain of cultured hippocampal neurons: implications for its involvement in mRNA transport. *J Neurosci* **19**(1): 288-297

Kim-Ha J, Smith JL, Macdonald PM (1991) oskar mRNA is localized to the posterior pole of the Drosophila oocyte. *Cell* **66**(1): 23-35

King RW, Glotzer M, Kirschner MW (1996) Mutagenic analysis of the destruction signal of mitotic cyclins and structural characterization of ubiquitinated intermediates. *Mol Biol Cell* **7**(9): 1343-1357

Knirr S, Breuer S, Paululat A, Renkawitz-Pohl R (1997) Somatic mesoderm differentiation and the development of a subset of pericardial cells depend on the not enough muscles (nem) locus, which contains the inscuteable gene and the intron located gene, skittles. *Mech Dev* **67**(1): 69-81

Knoblich JA, Jan LY, Jan YN (1995) Asymmetric segregation of Numb and Prospero during cell division. *Nature* **377**(6550): 624-627

Knoblich JA, Sauer K, Jones L, Richardson H, Saint R, Lehner CF (1994) Cyclin E controls S phase progression and its down-regulation during Drosophila embryogenesis is required for the arrest of cell proliferation. *Cell* **77**(1): 107-120

Krzemien J, Dubois L, Makki R, Meister M, Vincent A, Crozatier M (2007) Control of blood cell homeostasis in Drosophila larvae by the posterior signalling centre. *Nature* **446**(7133): 325-328

Kurucz E, Ando I, Sumegi M, Holzl H, Kapelari B, Baumeister W, Udvardy A (2002) Assembly of the Drosophila 26 S proteasome is accompanied by extensive subunit rearrangements. *Biochem J* **365**(Pt 2): 527-536

Lambert JD, Nagy LM (2002) Asymmetric inheritance of centrosomally localized mRNAs during embryonic cleavages. *Nature* **420**(6916): 682-686

Landgraf M, Bossing T, Technau GM, Bate M (1997) The origin, location, and projections of the embryonic abdominal motorneurons of Drosophila. *J Neurosci* **17**(24): 9642-9655

Lane ME, Sauer K, Wallace K, Jan YN, Lehner CF, Vaessin H (1996) Dacapo, a cyclin-dependent kinase inhibitor, stops cell proliferation during Drosophila development. *Cell* **87**(7): 1225-1235

Lasko P (2000) The drosophila melanogaster genome: translation factors and RNA binding proteins. *J Cell Biol* **150**(2): F51-56

Lawrence JB, Singer RH (1986) Intracellular localization of messenger RNAs for cytoskeletal proteins. *Cell* **45**(3): 407-415

Le Borgne R, Bardin A, Schweisguth F (2005) The roles of receptor and ligand endocytosis in regulating Notch signaling. *Development* **132**(8): 1751-1762

Lecuyer E, Yoshida H, Parthasarathy N, Alm C, Babak T, Cerovina T, Hughes TR, Tomancak P, Krause HM (2007) Global analysis of mRNA localization reveals a prominent role in organizing cellular architecture and function. *Cell* **131**(1): 174-187

Lee CY, Andersen RO, Cabernard C, Manning L, Tran KD, Lanskey MJ, Bashirullah A, Doe CQ (2006a) Drosophila Aurora-A kinase inhibits neuroblast self-renewal by regulating aPKC/Numb cortical polarity and spindle orientation. *Genes Dev* **20**(24): 3464-3474

Lee CY, Wilkinson BD, Siegrist SE, Wharton RP, Doe CQ (2006b) Brat is a Miranda cargo protein that promotes neuronal differentiation and inhibits neuroblast self-renewal. *Dev Cell* **10**(4): 441-449

Lee TV, Ding T, Chen Z, Rajendran V, Scherr H, Lackey M, Bolduc C, Bergmann A (2008) The E1 ubiquitin-activating enzyme Uba1 in Drosophila controls apoptosis autonomously and tissue growth non-autonomously. *Development* **135**(1): 43-52

Lehmann R. JR, Dietrich V. and Campos-Ortega J.A. (1983) On the phenotype and development of mutants of early neurogenesis in *Drosophila melanogaster*. **192**;62-74

Li L, Vaessin H (2000) Pan-neural Prospero terminates cell proliferation during Drosophila neurogenesis. *Genes Dev* **14**(2): 147-151

Li P, Yang X, Wasser M, Cai Y, Chia W (1997) Inscuteable and Staufen mediate asymmetric localization and segregation of prospero RNA during Drosophila neuroblast cell divisions. *Cell* **90**(3): 437-447

Lieber T, Wesley CS, Alcamo E, Hassel B, Krane JF, Campos-Ortega JA, Young MW (1992) Single amino acid substitutions in EGF-like elements of Notch and Delta modify Drosophila development and affect cell adhesion in vitro. *Neuron* **9**(5): 847-859

Liu H, Mardahl-Dumesnil M, Sweeney ST, O'Kane CJ, Bernstein SI (2003) Drosophila paramyosin is important for myoblast fusion and essential for myofibril formation. *J Cell Biol* **160**(6): 899-908

Liu TH, Li L, Vaessin H (2002) Transcription of the Drosophila CKI gene dacapo is regulated by a modular array of cis-regulatory sequences. *Mech Dev* **112**(1-2): 25-36

Long RM, Singer RH, Meng X, Gonzalez I, Nasmyth K, Jansen RP (1997) Mating type switching in yeast controlled by asymmetric localization of ASH1 mRNA. *Science* **277**(5324): 383-387

Lu B, Rothenberg M, Jan LY, Jan YN (1998) Partner of Numb colocalizes with Numb during mitosis and directs Numb asymmetric localization in Drosophila neural and muscle progenitors. *Cell* **95**(2): 225-235

MacDougall N, Clark A, MacDougall E, Davis I (2003) Drosophila gurken (TGFalpha) mRNA localizes as particles that move within the oocyte in two dynein-dependent steps. *Dev Cell* **4**(3): 307-319

Macieira-Coelho (2007) Asymmetric Cell Division. *Progress in Molecular and Subcellular Biology* **45**

Mandal L, Martinez-Agosto JA, Evans CJ, Hartenstein V, Banerjee U (2007) A Hedgehog- and Antennapedia-dependent niche maintains Drosophila haematopoietic precursors. *Nature* **446**(7133): 320-324

Marion RM, Fortes P, Beloso A, Dotti C, Ortin J (1999) A human sequence homologue of Staufen is an RNA-binding protein that is associated with polysomes and localizes to the rough endoplasmic reticulum. *Mol Cell Biol* **19**(3): 2212-2219

Matsuzaki F, Ohshiro T, Ikeshima-Kataoka H, Izumi H (1998) miranda localizes staufen and prospero asymmetrically in mitotic neuroblasts and epithelial cells in early Drosophila embryogenesis. *Development* **125**(20): 4089-4098

Maynard JC, Spana, E., Nicchitta, C.V. (2008) An Essential Role for Gp93, the Endoplasmic Reticulum Hsp90 Chaperone, in Growth Control. *A Dros Res Conf 49 : 163A*

Melton DA (1987) Translocation of a localized maternal mRNA to the vegetal pole of Xenopus oocytes. *Nature* **328**(6125): 80-82

Meraldi P, Honda R, Nigg EA (2004) Aurora kinases link chromosome segregation and cell division to cancer susceptibility. *Curr Opin Genet Dev* **14**(1): 29-36

Mettler U, Vogler G, Urban J (2006) Timing of identity: spatiotemporal regulation of hunchback in neuroblast lineages of Drosophila by Seven-up and Prospero. *Development* **133**(3): 429-437

Meyer CA, Kramer I, Dittrich R, Marzodko S, Emmerich J, Lehner CF (2002) Drosophila p27Dacapo expression during embryogenesis is controlled by a complex regulatory region independent of cell cycle progression. *Development* **129**(2): 319-328

Micchelli CA, Perrimon N (2006) Evidence that stem cells reside in the adult Drosophila midgut epithelium. *Nature* **439**(7075): 475-479

Micklem DR, Adams J, Grunert S, St Johnston D (2000) Distinct roles of two conserved Staufen domains in oskar mRNA localization and translation. *EMBO J* **19**(6): 1366-1377

Mingle LA, Okuhama NN, Shi J, Singer RH, Condeelis J, Liu G (2005) Localization of all seven messenger RNAs for the actin-polymerization nucleator Arp2/3 complex in the protrusions of fibroblasts. *J Cell Sci* **118**(Pt 11): 2425-2433

Mollinari C, Lange B, Gonzalez C (2002) Miranda, a protein involved in neuroblast asymmetric division, is associated with embryonic centrosomes of Drosophila melanogaster. *Biol Cell* **94**(1): 1-13

Morrison SJ, Shah NM, Anderson DJ (1997) Regulatory mechanisms in stem cell biology. *Cell* **88**(3): 287-298

Neuman-Silberberg FS, Schupbach T (1993) The Drosophila dorsoventral patterning gene gurken produces a dorsally localized RNA and encodes a TGF alpha-like protein. *Cell* **75**(1): 165-174

Nishikura K, Yoo C, Kim U, Murray JM, Estes PA, Cash FE, Liebhaber SA (1991) Substrate specificity of the dsRNA unwinding/modifying activity. *EMBO J* **10**(11): 3523-3532

Ohlstein B, Spradling A (2006) The adult Drosophila posterior midgut is maintained by pluripotent stem cells. *Nature* **439**(7075): 470-474

Ohlstein B, Spradling A (2007) Multipotent Drosophila intestinal stem cells specify daughter cell fates by differential notch signaling. *Science* **315**(5814): 988-992

Parmentier ML, Woods D, Greig S, Phan PG, Radovic A, Bryant P, O'Kane CJ (2000) Rapsynoid/partner of inscuteable controls asymmetric division of larval neuroblasts in Drosophila. *J Neurosci* **20**(14): RC84

Pesin JA, Orr-Weaver TL (2008) Regulation of APC/C Activators in Mitosis and Meiosis. *Annu Rev Cell Dev Biol* **24**: 475-499

Petritsch C, Tavosanis G, Turck CW, Jan LY, Jan YN (2003) The Drosophila myosin VI Jaguar is required for basal protein targeting and correct spindle orientation in mitotic neuroblasts. *Dev Cell* **4**(2): 273-281

Polson AG, Bass BL (1994) Preferential selection of adenosines for modification by double-stranded RNA adenosine deaminase. *EMBO J* **13**(23): 5701-5711

Prokop A, Bray S, Harrison E, Technau GM (1998) Homeotic regulation of segment-specific differences in neuroblast numbers and proliferation in the Drosophila central nervous system. *Mech Dev* **74**(1-2): 99-110

Prokop A, Technau GM (1991) The origin of postembryonic neuroblasts in the ventral nerve cord of Drosophila melanogaster. *Development* **111**(1): 79-88

Raff JW, Whitfield WG, Glover DM (1990) Two distinct mechanisms localise cyclin B transcripts in syncytial Drosophila embryos. *Development* **110**(4): 1249-1261

Rhyu MS, Jan LY, Jan YN (1994) Asymmetric distribution of numb protein during division of the sensory organ precursor cell confers distinct fates to daughter cells. *Cell* **76**(3): 477-491

Rivera-Pomar R, Niessing D, Schmidt-Ott U, Gehring WJ, Jackle H (1996) RNA binding and translational suppression by bicoid. *Nature* **379**(6567): 746-749

Roegiers F, Jan YN (2000) Staufen: a common component of mRNA transport in oocytes and neurons? *Trends Cell Biol* **10**(6): 220-224

Saunders C, Cohen RS (1999) The role of oocyte transcription, the 5'UTR, and translation repression and derepression in Drosophila gurken mRNA and protein localization. *Mol Cell* **3**(1): 43-54

Scadden AD (2005) The RISC subunit Tudor-SN binds to hyper-edited double-stranded RNA and promotes its cleavage. *Nat Struct Mol Biol* **12**(6): 489-496

Schaefer M, Shevchenko A, Knoblich JA (2000) A protein complex containing Inscuteable and the Galpha-binding protein Pins orients asymmetric cell divisions in Drosophila. *Curr Biol* **10**(7): 353-362

Schmid A, Chiba A, Doe CQ (1999) Clonal analysis of Drosophila embryonic neuroblasts: neural cell types, axon projections and muscle targets. *Development* **126**(21): 4653-4689

Schmidt H, Rickert C, Bossing T, Vef O, Urban J, Technau GM (1997) The embryonic central nervous system lineages of Drosophila melanogaster. II. Neuroblast lineages derived from the dorsal part of the neuroectoderm. *Dev Biol* **189**(2): 186-204

Schober M, Schaefer M, Knoblich JA (1999) Bazooka recruits Inscuteable to orient asymmetric cell divisions in Drosophila neuroblasts. *Nature* **402**(6761): 548-551

Schroeder T (2007) Asymmetric Cell Division in Normal and Malignant Hematopoietic Precursor Cells. *Cell Stem Cell* **1**(5): 479-481

Schuldt AJ, Adams JH, Davidson CM, Micklem DR, Haseloff J, St Johnston D, Brand AH (1998) Miranda mediates asymmetric protein and RNA localization in the developing nervous system. *Genes Dev* **12**(12): 1847-1857

Schweisguth F (2004) Regulation of notch signaling activity. *Curr Biol* **14**(3): R129-138

Seshaiah P, Andrew DJ (1999) WRS-85D: A tryptophanyl-tRNA synthetase expressed to high levels in the developing Drosophila salivary gland. *Mol Biol Cell* **10**(5): 1595-1608

Seyit G, Rockel B, Baumeister W, Peters J (2006) Size matters for the tripeptidylpeptidase II complex from Drosophila: The 6-MDa spindle form stabilizes the activated state. *J Biol Chem* **281**(35): 25723-25733

Shen CP, Jan LY, Jan YN (1997) Miranda is required for the asymmetric localization of Prospero during mitosis in Drosophila. *Cell* **90**(3): 449-458

Shen CP, Knoblich JA, Chan YM, Jiang MM, Jan LY, Jan YN (1998) Miranda as a multidomain adapter linking apically localized Inscuteable and basally localized Staufen and Prospero during asymmetric cell division in Drosophila. *Genes Dev* **12**(12): 1837-1846

Simmonds AJ, dosSantos G, Livne-Bar I, Krause HM (2001) Apical localization of wingless transcripts is required for wingless signaling. *Cell* **105**(2): 197-207

Singh SR, Liu W, Hou SX (2007) The adult Drosophila malpighian tubules are maintained by multipotent stem cells. *Cell Stem Cell* **1**(2): 191-203

Skeath JB, Thor S (2003) Genetic control of Drosophila nerve cord development. *Curr Opin Neurobiol* **13**(1): 8-15

Slack C, Overton PM, Tuxworth RI, Chia W (2007) Asymmetric localisation of Miranda and its cargo proteins during neuroblast division requires the anaphase-promoting complex/cyclosome. *Development* **134**(21): 3781-3787

Somers WG, Saint R (2003) A RhoGEF and Rho family GTPase-activating protein complex links the contractile ring to cortical microtubules at the onset of cytokinesis. *Dev Cell* **4**(1): 29-39

Song X, Zhu CH, Doan C, Xie T (2002) Germline stem cells anchored by adherens junctions in the Drosophila ovary niches. *Science* **296**(5574): 1855-1857

Spana EP, Doe CQ (1995) The prospero transcription factor is asymmetrically localized to the cell cortex during neuroblast mitosis in Drosophila. *Development* **121**(10): 3187-3195

St Johnston D, Beuchle D, Nusslein-Volhard C (1991) Staufen, a gene required to localize maternal RNAs in the Drosophila egg. *Cell* **66**(1): 51-63

St Johnston D, Brown NH, Gall JG, Jantsch M (1992) A conserved double-stranded RNA-binding domain. *Proc Natl Acad Sci U S A* **89**(22): 10979-10983

St Johnston D, Driever W, Berleth T, Richstein S, Nusslein-Volhard C (1989) Multiple steps in the localization of bicoid RNA to the anterior pole of the Drosophila oocyte. *Development* **107 Suppl:** 13-19

Steneberg P, Englund C, Kronhamn J, Weaver TA, Samakovlis C (1998) Translational readthrough in the hdc mRNA generates a novel branching inhibitor in the drosophila trachea. *Genes Dev* **12**(7): 956-967

Steneberg P, Samakovlis C (2001) A novel stop codon readthrough mechanism produces functional Headcase protein in Drosophila trachea. *EMBO Rep* **2**(7): 593-597

Struhl G (1989) Differing strategies for organizing anterior and posterior body pattern in Drosophila embryos. *Nature* **338**(6218): 741-744

Stuttem I, Campos-Ortega JA (1991) Cell commitment and cell interactions in the ectoderm of Drosophila melanogaster. *Development* **Suppl 2:** 39-46

Taghert PH, Doe CQ, Goodman CS (1984) Cell determination and regulation during development of neuroblasts and neurones in grasshopper embryo. *Nature* **307**(5947): 163-165

Takizawa PA, Sil A, Swedlow JR, Herskowitz I, Vale RD (1997) Actin-dependent localization of an RNA encoding a cell-fate determinant in yeast. *Nature* **389**(6646): 90-93

Thio GL, Ray RP, Barcelo G, Schupbach T (2000) Localization of gurken RNA in Drosophila oogenesis requires elements in the 5' and 3' regions of the transcript. *Dev Biol* **221**(2): 435-446

Tio M, Udolph G, Yang X, Chia W (2001) cdc2 links the Drosophila cell cycle and asymmetric division machineries. *Nature* **409**(6823): 1063-1067

Truman JW, Bate M (1988) Spatial and temporal patterns of neurogenesis in the central nervous system of Drosophila melanogaster. *Dev Biol* **125**(1): 145-157

Uv AE, Harrison EJ, Bray SJ (1997) Tissue-specific splicing and functions of the Drosophila transcription factor Grainyhead. *Mol Cell Biol* **17**(11): 6727-6735

van de Weerdt BC, Medema RH (2006) Polo-like kinases: a team in control of the division. *Cell Cycle* **5**(8): 853-864

Wang C, Dickinson LK, Lehmann R (1994) Genetics of nanos localization in Drosophila. *Dev Dyn* **199**(2): 103-115

Wang C, Lehmann R (1991) Nanos is the localized posterior determinant in Drosophila. *Cell* **66**(4): 637-647

Wang H, Chia W (2005) Drosophila neural progenitor polarity and asymmetric division. *Biol Cell* **97**(1): 63-74

Wang H, Ouyang Y, Somers WG, Chia W, Lu B (2007) Polo inhibits progenitor self-renewal and regulates Numb asymmetry by phosphorylating Pon. *Nature* **449**(7158): 96-100

Wang H, Somers GW, Bashirullah A, Heberlein U, Yu F, Chia W (2006) Aurora-A acts as a tumor suppressor and regulates self-renewal of Drosophila neuroblasts. *Genes Dev* **20**(24): 3453-3463

Weaver TA, White RA (1995) headcase, an imaginal specific gene required for adult morphogenesis in Drosophila melanogaster. *Development* **121**(12): 4149-4160

White K, Grether ME, Abrams JM, Young L, Farrell K, Steller H (1994) Genetic control of programmed cell death in Drosophila. *Science* **264**(5159): 677-683

White K, Kankel DR (1978) Patterns of cell division and cell movement in the formation of the imaginal nervous system in Drosophila melanogaster. *Dev Biol* **65**(2): 296-321

Wickham L, Duchaine T, Luo M, Nabi IR, DesGroseillers L (1999) Mammalian staufen is a double-stranded-RNA- and tubulin-binding protein which localizes to the rough endoplasmic reticulum. *Mol Cell Biol* **19**(3): 2220-2230

Wilson DM, 3rd, Deutsch WA, Kelley MR (1994) Drosophila ribosomal protein S3 contains an activity that cleaves DNA at apurinic/apyrimidinic sites. *J Biol Chem* **269**(41): 25359-25364

Wodarz A, Huttner WB (2003) Asymmetric cell division during neurogenesis in Drosophila and vertebrates. *Mech Dev* **120**(11): 1297-1309

Wu M, Kwon HY, Rattis F, Blum J, Zhao C, Ashkenazi R, Jackson TL, Gaiano N, Oliver T, Reya T (2007) Imaging hematopoietic precursor division in real time. *Cell Stem Cell* **1**(5): 541-554

Yamamoto A, Guacci V, Koshland D (1996) Pds1p is required for faithful execution of anaphase in the yeast, Saccharomyces cerevisiae. *J Cell Biol* **133**(1): 85-97

Yoon YJ, Mowry KL (2004) Xenopus Staufen is a component of a ribonucleoprotein complex containing Vg1 RNA and kinesin. *Development* **131**(13): 3035-3045

Young JC, Moarefi I, Hartl FU (2001) Hsp90: a specialized but essential protein-folding tool. *J Cell Biol* **154**(2): 267-273

Yousef MS, Kamikubo H, Kataoka M, Kato R, Wakatsuki S (2008) Miranda cargo-binding domain forms an elongated coiled-coil homodimer in solution: implications for asymmetric cell division in Drosophila. *Protein Sci* **17**(5): 908-917

Yu F, Morin X, Cai Y, Yang X, Chia W (2000) Analysis of partner of inscuteable, a novel player of Drosophila asymmetric divisions, reveals two distinct steps in inscuteable apical localization. *Cell* **100**(4): 399-409

Zhang HL, Eom T, Oleynikov Y, Shenoy SM, Liebelt DA, Dictenberg JB, Singer RH, Bassell GJ (2001) Neurotrophin-induced transport of a beta-actin mRNP complex increases beta-actin levels and stimulates growth cone motility. *Neuron* **31**(2): 261-275

Zhang J, Houston DW, King ML, Payne C, Wylie C, Heasman J (1998) The role of maternal VegT in establishing the primary germ layers in Xenopus embryos. *Cell* **94**(4): 515-524

Die VDM Verlagsservicegesellschaft sucht für wissenschaftliche Verlage abgeschlossene und herausragende

Dissertationen, Habilitationen, Diplomarbeiten, Master Theses, Magisterarbeiten usw.

für die kostenlose Publikation als Fachbuch.

Sie verfügen über eine Arbeit, die hohen inhaltlichen und formalen Ansprüchen genügt, und haben Interesse an einer honorarvergüteten Publikation?

Dann senden Sie bitte erste Informationen über sich und Ihre Arbeit per Email an *info@vdm-vsg.de*.

Sie erhalten kurzfristig unser Feedback!

VDM Verlagsservicegesellschaft mbH
Dudweiler Landstr. 99
D - 66123 Saarbrücken

Telefon +49 681 3720 174
Fax +49 681 3720 1749

www.vdm-vsg.de

Die VDM Verlagsservicegesellschaft mbH vertritt

Printed by Books on Demand GmbH, Norderstedt / Germany